THE LOGIC OF
LEVIATHAN

THE LOGIC OF LEVIATHAN

THE MORAL AND POLITICAL THEORY
OF THOMAS HOBBES

BY

DAVID P. GAUTHIER

Professor of Philosophy
University of Toronto

OXFORD
AT THE CLARENDON PRESS

Oxford University Press, Walton Street, Oxford OX2 6DP

OXFORD LONDON GLASGOW
NEW YORK TORONTO MELBOURNE WELLINGTON
KUALA LUMPUR SINGAPORE JAKARTA HONG KONG TOKYO
DELHI BOMBAY CALCUTTA MADRAS KARACHI
NAIROBI DAR ES SALAAM CAPE TOWN

ISBN 0 19 824616 1

© Oxford University Press 1969

First published 1969
Reprinted in paperback 1979

Printed in Great Britain
at the University Press, Oxford
by Eric Buckley
Printer to the University

PREFACE

THIS study presents what I consider the most plausible reading of Hobbes's moral and political theory. The letter of Hobbes's text is neither fully consistent nor fully unambiguous. I have therefore based my interpretation on three considerations: I have tried to present a theory which is coherent, plausible, and true to what I find to be the spirit of Hobbes's thought. I have used the English works exclusively in this task; I do not pretend to Latin scholarship, and I have done no more than check my reading against the Latin texts, to assure myself that they do not present objections to it.

This study also contains some minor expansions of, and substantial critical reflections on, Hobbes's theory. Occasionally Hobbes's arguments fail to establish his point. I have sometimes found it possible to construct arguments, coherent with the main lines of his theory, which would establish his point, and I have included those arguments—always sharply distinguished from my presentation of Hobbes's actual position. And I have also tried to present general assessments of both the moral and the political theory, for it seems to me that our interest in Hobbes is not primarily in what he said, but in what we can accept and use of what he said.

I make no apology for my extensive use of direct quotation. My aim has been to give the reader the evidence on which I base my interpretation and criticism, and to enable him to read this study without constantly referring to Hobbes's own works.

My interest in Hobbes began in 1958, when I first taught a course in seventeenth- and eighteenth-century moral and political philosophy at the University of Toronto. At that time I became acquainted with Howard Warrender's study, *The Political Philosophy of Hobbes*, which convinced me both that Hobbes could be a profitable subject for philosophical reflection, and that Warrender's own reflections, although both ingenious and illuminating, could not be accepted.

In 1960 I wrote a brief, unpublished paper on Hobbes's moral theory. Gradually I became aware that a larger project was taking shape, and in the summers of 1963 and 1964 I wrote a first draft of this present work. The present draft, in which almost all of the apparatus is introduced, was written in the spring of 1967 at Oxford.

Why write on Hobbes? Hobbes constructs a political theory which bases unlimited political authority on unlimited individualism. The conclusion requires the premiss; anything less than unlimited individualism would justify only limited political authority. But the premiss is too strong for the conclusions; as I attempt to show, from unlimited individualism only anarchy follows. The theory is a failure.

But it has two outstanding merits. First of all, Hobbes introduces a number of important moral and political concepts which deserve our attention. Obligation is his basic moral concept, and I know no other philosopher who is both so illuminatingly right and so illuminatingly wrong about it. Authorization is his basic political concept, and I know no other political theorist who has recognized and exploited it so ably in interpreting the relation between sovereign and subject, or if you like, government and citizen.

And secondly, Hobbes has much to teach those of us who would base limited political authority on limited individualism. He relies neither on the goodwill of men—their willingness to consider each other's interests for their own sake, and not as means to self-satisfaction, nor on the efficacy of institutions, as the means of both concentrating and limiting political power.

His refusal to rely on either of these explains the failure of his theory. But if we look upon his argument as a limiting case, an attempt to construct a political order on the least favourable assumptions, we shall appreciate its value. For in understanding its failure, we also understand what must be added to overcome that failure. We can then argue from a minimum reliance on human goodwill and institutional efficacy. And this is surely to provide one reasonable task for constructive political theory—to show the minimal assumptions on which a political order may be constructed.

All quotations from Hobbes are taken from *The English Works of Thomas Hobbes*, ed. Sir William Molesworth, 11 vols., London, 1839–45. '*E.W.* iii' abbreviates '*English Works*, vol. iii'.

I am grateful to the University of Toronto for providing a Humanities Research Grant to enable me to spend the spring of 1967 in Oxford. I am also grateful to St. Antony's College, Oxford, for electing me to Senior Associate Membership for Trinity Term, 1967, and for arranging room and board.

D. P. G.

University of Toronto
June 1968

CONTENTS

I

THE NATURE OF MAN

To understand morals and politics, understand man. Leviathan, 'that *mortal god*, to which we owe under the *immortal God*, our peace and defence' (*E.W.* iii, p. 158), the body politic which gives substance to our moral conceptions, is the product of human reason reflecting on human passion.

That Hobbes's psychology is essential to his moral and political theories is one of the main theses of our argument. This view has been challenged by many recent interpreters and critics.[1] But we must begin, not with a discussion about Hobbes's argument, but with a statement of that argument.

Our first task is to uncover Hobbes's conception of his own method. This will show why he considers psychology to be the necessary foundation of moral and political science, and so will lead us into his account of the nature of man.

1. METHOD

The Introduction to *Leviathan* compares the making of Commonwealth by the art of man with the making of man by the art of nature. Nature makes life, which is 'but a motion of limbs, the beginning whereof is in some principal part within' (*E.W.* iii, p. ix); man makes automata, self-moving engines and machines, which have an artificial life. Man is the most excellent work of nature; Commonwealth is the most excellent work of man.

This account suggests a naturalistic explanation of human mechanical art. But the direction of fit is in fact the reverse.

---

[1] The challenge appears in its boldest form in A. E. Taylor, 'The Ethical Doctrine of Hobbes', *Philosophy*, vol. xiii, 1938, pp. 406–24, reprinted in Keith C. Brown (ed.), *Hobbes Studies*, Oxford, 1965, pp. 35–55. It is fully developed in Howard Warrender, *The Political Philosophy of Hobbes*, Oxford, 1957. It has been, of course, the subject of much discussion elsewhere.

'For what is the *heart*, but a *spring*; and the *nerves*, but so many *strings*; and the *joints*, but so many *wheels*, giving motion to the whole body, . . . ?' (*E.W.* iii, p. ix). It is nature which imitates human artifice; God, the author of nature, is the divine mechanic.

Hobbes is a methodological mechanist. He seeks to construct a unified science, proceeding from a study of body in general to a study of that particular body, man, and then to a study of man-made artificial bodies. But 'the causes of the motions of the mind are known, not only by ratiocination, but also by the experience of every man that takes the pains to observe those motions within himself' (*E.W.* i, p. 73). Since the motions of the mind sufficiently explain human action, and hence the making of artificial bodies, it follows that the study of that artificial body, the Commonwealth, 'grounded on its own principles sufficiently known by experience', does 'not stand in need of the former sections' (*E.W.* ii, p. xx).

What unifies Hobbes's philosophy, relating his study of politics to his physics, is in fact not the thread of deductive materialism, but the stitch of mechanical explanation. And the name of this stitch is Motion—for motion is Hobbes's conceptual key to the understanding of all reality.

Hobbes illustrates the application of the principles of mechanical explanation to political bodies in the Preface to *De Cive*:

Concerning my method, I thought it not sufficient to use a plain and evident style in what I have to deliver, except I took my beginning from the very matter of civil government, and thence proceeded to its generation and form, and the first beginning of justice. For everything is best understood by its constitutive causes. For as in a watch, or some such small engine, the matter, figure, and motion of the wheels cannot well be known, except it be taken insunder and viewed in parts; so to make a more curious search into the rights of states and duties of subjects, it is necessary, I say, not to take them insunder, but yet that they be so considered as if they were dissolved; that is, that we rightly understand what the quality of human nature is, in what matters it is, in what not, fit to make up a civil government, and how men must be agreed amongst themselves that intend to grow up into a well-grounded state. (*E.W.* ii, p. xiv)

This is the resoluto-compositive method of Galileo, developed in the School of Padua. We resolve, or as Hobbes often
prefers to say, analyse a whole—a watch, or a Commonwealth
—into its constituent parts, discovering the relations among
the parts which then enables us to recompose—or synthesize
—the whole. Had Hobbes succeeded in his grand design for
a unified science, questions of morals and politics would be
treated in a purely synthetic manner, by derivation from the
supreme relations or principles established through an analysis
of body. But in his political writings experience replaces this
derivation. The maxim *'nosce teipsum, read thyself'* (*E.W.* iii,
p. xi) provides the analytic moment sufficient to establish the
principles of moral and political philosophy.

Hobbes illustrates the moments of resolution and composition in the solution of a political problem in a brief but
illuminating passage of *De Corpore* which suggests much of his
positive political doctrine:

> For if a question be propounded, as, *whether such an action be just or
> unjust*; if that *unjust* be resolved into *fact against law*, and that notion
> *law* into the *command* of him or them that have *coercive power*; and that
> *power* be derived from the *wills* of men that constitute such power, to
> the end they may live in peace, they may at last come to this, that the
> appetites of men and the passions of their minds are such, that, unless
> they be restrained by some power, they will always be making war upon
> one another; which may be known to be so by any man's experience,
> that will but examine his own mind. (*E.W.* i, p. 74)

Thus 'to describe the nature of this artificial man'—of
Commonwealth—'I will consider
'First, the *matter* thereof, and the *artificer*; both which is
man' (*E.W.* iii, p. x).

Hobbes's methodological mechanism presupposes the essential similarity of physics and politics. Both are sciences which
explain phenomena in terms of consequential relationships,
which Hobbes seems to have taken indifferently as causal
connections or logical entailments. The political scientist
enables us to understand Commonwealth, to know its constituent causes, and so—for knowledge is power—to enable

men to erect a Commonwealth should its causes be in their control.

But is Hobbes's aim to provide understanding—even that understanding which is power? An immediate objection to Hobbes's treatment of politics is raised by Richard Peters:

For was he really proposing to explain the causes of a state at all? When he says that he will enquire about the matters in which human nature is *fit* to make up a civil government and 'how men *must* be agreed amongst themselves that intend to grow up into a *well-grounded* state', he obviously is proposing to show how a rational state *ought* to be constructed; he surely is not attempting an explanation of actual states. However, he proceeds as if he were conducting a Galilean experiment rather than enunciating a rational plan for reconstruction.[1]

Suppose it be granted that Hobbes can explain how Commonwealth functions in terms of its constituent causes. This, it is urged, does not tell him what he wants to know—how Commonwealth ought to function. I may take a watch apart, examine its parts and their relationship, and so determine how to put it back together. But nothing in this inquiry reveals to me whether the watch is functioning correctly, whether it is keeping time. Indeed, nothing in my inquiry shows that the function of the watch is to keep time.

However, if I do know the function of the watch, then by taking it apart and ascertaining the manner of its working I can find out how to reconstruct it so that it will keep correct time. The Galilean experiment, combined with my prior knowledge of the function of the watch, does enable me to determine what to do. Hobbes's mistake, then, is not in seeking to understand Commonwealth by means of a Galilean experiment, but in supposing that this understanding is sufficient to determine a rational plan for the reconstruction of Commonwealth. Prior knowledge of the function of Commonwealth is needed. If I know what a well-grounded state is, then I can determine, by a Galilean experiment, those qualities in human nature which fit man for life in such a state, and how men must be agreed in order to construct it.

[1] Richard Peters, *Hobbes*, Harmondsworth, 1956, pp. 74–5.

Thus, even if Peters's objection be allowed, Hobbes can still justify his method, and his initial concern with the nature of man. Peters is not challenging the relevance of Hobbes's psychology to his moral and political theories. His essential complaint is that Hobbes commits a logical fallacy in seeking to derive a normative conclusion, about the proper construction of the state, from purely factual premisses about the nature of man. We shall be able to determine whether Hobbes is guilty of such a fallacy when we have considered these premisses.

2. MAN

Since Hobbes understands all phenomena in terms of motion, living beings can be defined only in terms of a motion common to all. As we have noted, in the Introduction to *Leviathan*, Hobbes treats life as 'but a motion of limbs', as animation. However, this is not his considered view, for elsewhere he groups together 'the *course* of the *blood*, the *pulse*, the *breathing*, the *concoction*, *nutrition*, *excretion*, &c.' (*E.W.* iii, p. 38) as vital motion, the generic motion of the living world. These motions are maintained in all living beings without interruption, from their generation to their death. Animation belongs properly to a second class of motions.

Hobbes does not provide, nor does he think it necessary to provide, a complete list of vital motions. Nor does he consider whether the examples of vital motion which he does provide do in fact possess the characteristics essential to vital motion. No doubt his conception of life as motion does have a basis in the observation of bodily processes which continue throughout life and cease in death. However, the existence of vital motion is properly a metaphysical necessity, a requirement of the mechanical framework which Hobbes employs in all his inquiries. That some particular motion is vital may be queried, but vital motion there must be.

All animals possess, in addition to vital motion, voluntary motion 'as to *go*, to *speak*, to *move* any of our limbs, in such manner as is first fancied in our minds' (ibid.).

Animation then defines, not the living world, but the animal world.

Human action is voluntary motion; human passion is the beginning of voluntary motion. There is no species of motion peculiar to man. It would seem to follow that there can be no essential difference between man and the other animals, since an essential difference can only be a difference in species of motion.

We may trace the origin of voluntary motion to sense perception—itself, of course, a type of motion in Hobbes's account. External objects act on the organs of sense, setting up a pressure directed inward toward the brain. There resistance occurs, an outward counter-pressure is set up, and this motion is sense. We suppose the objects of sense to be without, because this motion is directed outward.

That which is in motion continues in motion until altered by some other force. Thus the motion of sense continues after the removal of the external object, and this continued motion, gradually diminished by other pressures, is imagination or fancy.

From the brain the motion of imagination proceeds to the heart, which is the seat of vital motion. Coming into contact with vital motion, it must either help or hinder it; 'when it *helpeth*, it is called *delight*, *contentment*, or *pleasure*, which is nothing really but motion about the heart, as conception is nothing but motion in the head: . . . but when such motion *weakeneth* or hindereth the vital motion, then it is called *pain*' (*E.W.* iv, p. 31).

Resistance in the heart directs the motion of imagination outward once more. This outward motion is the first beginning of voluntary motion, called *endeavour*.[1] In so far as the motion of imagination helps vital motion, endeavour is directed toward the object which is the cause of imagination, and so is called appetite or desire. In so far as the motion of imagination

[1] The importance of endeavour in Hobbes's philosophic system as a whole can not concern us here. An admirable treatment of this concept is found in J. W. N. Watkins, *Hobbes's System of Ideas*, London, 1965, pp. 123 ff.

hinders vital motion, endeavour is directed away from the object, and is called aversion.

The several passions of man are species of desire and aversion. Thus action, which consists in voluntary motion, arises from the passions, internal motions of desire and aversion. And this action is directed toward those objects whose effects enhance vital motion, and away from those objects whose effects impede vital motion. Men—and indeed all animals—are thus conceived by Hobbes as self-maintaining mechanisms—engines whose motion is such that it enables them to continue to move as long as continued motion is possible.

From this account of vital and voluntary motion, it follows that each man seeks, and seeks only, to preserve and to strengthen himself. A concern for continued well-being is both the necessary and the sufficient ground of human action. Hence man is necessarily selfish.

Before continuing our sketch of human nature it will be well to pause, as Hobbes does, and introduce the basic terms of evaluation. Hobbes treats the definitions of these terms as an integral part of his psychology. An examination of these definitions is, therefore, of the first importance to an understanding of the relation between his psychology and his ethics and politics. Hobbes says:

> But whatsoever is the object of any man's appetite or desire, that is it which he for his part calleth *good*: and the object of his hate and aversion, *evil*; and of his contempt, *vile* and *inconsiderable*. For these words of good, evil, and contemptible, are ever used with relation to the person that useth them: there being nothing simply and absolutely so; nor any common rule of good and evil, to be taken from the nature of the objects themselves; but from the person of the man, ... (*E.W.* iii, p. 41)

It is essential to note that this account, although it defines evaluative terms psychologically, does not define them in terms of Hobbes's particular psychological theory. 'This is good' means 'this is an object of desire', not 'this is an object enhancing vital motion'. If as a matter of fact we desire what enhances vital motion, then what is good will enhance vital motion. But this conclusion does not follow merely from the definition of 'good'.

I propose to introduce the following terminological distinction. I shall say that, for Hobbes, the *formal* meaning of 'good' is conveyed by the equivalence: 'this is good' = 'this is an object of desire'. The *material* meaning of 'good' is conveyed by the equivalence: 'this is good' M = 'this is an object enhancing vital motion'. The rationale for introducing this terminology will be apparent only when we turn to Hobbes's moral theory. But to anticipate, we shall find that a failure to distinguish formal and material meanings of several of Hobbes's basic concepts is one of the principal sources of the dispute about the relevance of his psychology to his ethics.

Thus Hobbes's account of good and evil is formally subjectivistic, but not selfish. Someone who considered man fundamentally altruistic in nature could accept the definitions. The measure of Hobbes's audacity as a moral and political thinker is not that he attempted to erect his structure on the subjectivist foundations of his formal definitions, but that he attempted it on the self-centred foundations of his material characterizations.

Endeavour, if unchecked, results in action. We seek to obtain what we call good and to avoid what we call evil. But it may be that

> to our first appetite there succeedeth some conception of evil to happen to us by such actions, which is fear, and which holdeth us from proceeding. And to that fear may succeed a new appetite, and to that appetite another fear alternately, till the action be either done, or some accident come between, to make it impossible. . . . This *alternate succession of appetite and fear* during all the time the action is in our power to do or not to do, is that we call *deliberation*; . . . (*E.W.* iv, pp. 67–8)

The last appetite or aversion in deliberation is called the *will*. The will is always directed toward the promotion of our well-being, as we conceive it. But deliberation, which is Hobbes's version of practical reasoning, is fallible. Not only may we mistake the means to our well-being, but passion may lead us to misconceive well-being itself. Hence the results of

our practical reasoning need not accord with reason, and so
Hobbes rejects the Scholastic definition of will as rational
appetite, '(f) or if it were, then could there be no voluntary
act against reason' (*E.W.* iii, p. 48).

In a passage which is not altogether free from confusion
Hobbes tells us:

Continual success in obtaining those things which a man from time to
time desireth, that is to say, continual prospering, is that men call
FELICITY; I mean the felicity of this life. For there is no such thing as
perpetual tranquillity of mind, while we live here; because life itself is
but motion, and can never be without desire, nor without fear, no more
than without sense. (*E.W.* iii, p. 51)

It does not follow from the fact that life is but motion—
vital motion—that there is no tranquillity in this life. Hobbes
accepts, by anticipation, Newton's first law that an object
continues in its state, whether of rest or uniform motion,
unless affected by some external force. Hence if vital motion
continues, neither enhanced nor hindered, tranquillity should
be attained.

But if we turn from vital to voluntary motion, we can see
what Hobbes's real claim is. For external bodies continually
affect the organs of sense; thus the motions of sense are continual, and ever varying. Imagination, then, which is decaying
sense, must also be continual and varying. But imagination, on
reaching the heart, gives rise to endeavour, and endeavour
results in voluntary action. Hence endeavour and action must
also be continual and varying, and so felicity consists in continual prospering, and not in perpetual tranquillity.

The self-maintaining engines are constantly active, seeking
ever new objects of desire, avoiding ever new objects of aversion. The means of success in this unceasing activity is
power, which Hobbes defines as a man's 'present means; to
obtain some future apparent good' (*E.W.* iii, p. 74).

By this power I mean the same with the faculties of the *body, nutritive,
generative, motive,* and of the *mind, knowledge*; and besides these, such
further power as by them is acquired, *viz. riches, place* of authority,

friendship or *favour*, and *good fortune*; *which* last is really nothing else but the favour of God Almighty. (*E.W.* iv, pp. 37–8)

In his earliest account of human nature Hobbes treats power as essentially a comparative quantity. 'And because the power of one man resisteth and hindereth the effects of the power of another, *power* simply is no more, but the *excess* of the power of one above that of another: . . .' (*E.W.* iv, p. 38). In *Leviathan* this reduction of power to *greater* power is only suggested by Hobbes's characterization of natural power as 'the eminence of the faculties of body, or mind' (*E.W.* iii, p. 74), but again the powers of different individuals are treated as essentially opposed. Hobbes has described *man*; now he turns to consider *men*, and his basic concern is the competition among them.

Why do Hobbes's self-maintaining engines necessarily compete? Before turning to his answer, we must elaborate one aspect of human nature which Hobbes does not treat systematically—man's reason. For rationality will prove to be the essential foundation of Hobbes's moral and political theory.

3. REASON

Hobbes conceives mental discourse, or thinking, in the same manner as deliberation. Alternate opinions succeed one another, instead of alternate appetites. Opinions are the first beginnings of the movement of imagination in the brain, just as endeavours are the first beginnings of the movement of imagination in the heart. 'And as the last appetite in deliberation, is called the *will*; so the last opinion in search of the truth of past, and future, is called the JUDGMENT, . . .' (*E.W.* iii, p. 52).

The Procrustean bed of Hobbes's uncompromising mechanism forces him to speak as if discourse consists merely in alternating opinions of the truth of some matter—'thoughts that the thing will be, and will not be; or that it has been, and has not been, alternately' (ibid.). But in fact he wishes to include several quite distinct varieties of 'alternation' as different ways in which one opinion may succeed to another.

Our concern is only with that type of 'alternation' in which the succession of opinions represents the successive stages in argument. Such discourse, which is reasoning, begins with the absolute knowledge of fact—that this is, or was—which is the direct province of sense and memory, and proceeds to a knowledge of consequences—'that if this be, that is; if this has been, that has been; if this shall be, that shall be' (ibid.), which we call Science.

Knowledge of fact, for Hobbes, is recorded in definitions. And so he says that when discourse 'begins with the definitions of words, and proceeds by connexion of the same into general affirmations, and of these again into syllogisms; the end or last sum is called the conclusion; and the thought of the mind by it signified, is that conditional knowledge, or knowledge of the consequence of words, which is commonly called SCIENCE' (*E.W.* iii, pp. 52–3). Reason, equally, is characterized in a literally nominalist manner, for it 'is nothing but *reckoning*, that is adding and subtracting, of the consequences of general names agreed upon for the *marking* and *signifying* of our thoughts' (*E.W.* iii, p. 30).

When Hobbes speaks of 'general names', he is of course not saying that we reason about general things, for his nominalism denies that there are such. Rather, he is reminding us that, although we reason about particular things, we do not reason about their particularity; proper names have no consequences. When I conclude from the appearance of the sky that it will probably rain, I am reasoning about today's sky and tomorrow's weather, but in terms of a connection between a type of sky and a type of weather.

The exercise of reason, then, provides us with science, and science is power. For it 'is the knowledge of consequences, and dependance of one fact upon another: by which, out of that we can presently do, we know how to do something else when we will, or the like another time' (*E.W.* iii, p. 35). In this way reason aids deliberation by showing us the means to obtain, or to avoid, what we will.

But reason serves deliberation in a further way. The

succession of appetites and aversions in deliberation reflects the succession of opinions in discourse. As we determine further consequences of some act, our desire to perform it, or not to perform it, alters. Our last opinion is judgement; our judgement of the several consequences of an action determines our last appetite, will. In so far as we judge correctly and fully, we will what conduces most to our ends. And so correct reasoning enables will to be rational appetite.

Hobbes does not develop this relationship between discourse and deliberation. He conceives them as analogues, but does not show explicitly how one serves the other. But this is surely implicit in his entire account of reason, and in his insistence upon the practical significance of science. Reason is an essential part of the equipment of the self-maintaining engine, man.

But reasoning, or discourse, like deliberation, is fallible— 'not but that reason itself is always right reason, . . . but no one man's reason, nor the reason of any one number of men, makes the certainty' (*E.W.* iii, pp. 30–1). For 'if the first ground of . . . discourse, be not definitions; or if the definitions be not rightly joined together into syllogisms, then the end or conclusion, is again OPINION, namely of the truth of somewhat said, though sometimes in absurd and senseless words, without possibility of being understood' (*E.W.* iii, p. 53).

If actual reasoning is always uncertain, how is reason itself —*right reason*—to be determined? The concept of right reason is of fundamental import to Hobbes's moral and political theory. Yet his treatment of it is neither complete nor consistent. In *Human Nature* he gives this characterization of it:

> Now when a man *reasoneth* from *principles* that are *found* indubitable by experience, all deceptions of sense and equivocation of words avoided, the conclusion he maketh is said to be *according to right reason*: but when from his conclusion a man may, by good ratiocination, derive that which is *contradictory* to any evident truth whatsoever, then he is said to have concluded *against reason*: and such a conclusion is called *absurdity*. (*E.W.* iv, p. 24)

The test of right reason would seem to be the consistency of the conclusion with established and evident truth. And this evidence is based on sense, 'for the truth of a proposition is never evident, until we conceive the meaning of the words or terms whereof it consisteth, which are always conceptions of the mind: nor can we remember those conceptions, without the thing that produced the same by our senses' (*E.W.* iv, p. 28).

But in *Leviathan* Hobbes presents a very different account:

And therefore, as when there is a controversy in an account, the parties must by their own accord, set up, for right reason, the reason of some arbitrator, or judge, to whose sentence they will both stand, or their controversy must either come to blows, or be undecided, for want of a right reason constituted by nature; so is it also in all debates of what kind soever. And when men that think themselves wiser than all others, clamour and demand right reason for judge, yet seek no more, but that things should be determined, by no other men's reason but their own, it is as intolerable in the society of men, as it is in play after trump is turned, to use for trump on every occasion, that suite whereof they have most in their hand. (*E.W.* iii, p. 31)

Right reason has no natural standard—no appeal to 'evident truth' will suffice. Rather, right reason must be established by convention. Just as no suit in cards is by nature trump, but only by the rules of the game, so no judgement is by nature according to right reason, but only by the rules of social intercourse. In the absence of society, each man must presumably judge for himself; *his* reason is, for *him*, right reason.

Conflicting theories of truth seem to underlie these two accounts of right reason. In *Human Nature* truth depends on an ultimate appeal to what is evident in terms of conceptions derived from sense; in *Leviathan* truth depends on an appeal to an agreed, conventional standard. Our concern is not with Hobbes's epistemology, and we cannot examine the problems which these theories raise. Rather, all we can do is to call attention to these opposed views of truth and of right reason, so that we may have them in mind when we turn to Hobbes's use of the concept of right reason in developing the bases of his ethics.

4. MEN

Hobbes's self-maintaining engines are forced into constant activity by the presentation of ever-new objects of desire and aversion in sense-experience. The world is full of helps and hindrances to their vital motions; they endeavour to secure the helps and to avoid the hindrances.

Power is the means of success in this unending activity. But no finite degree of power can ever ensure success. The hindrances may prove too great, the helps too small. Vital motion may, and eventually must, be stilled. Thus when Hobbes comes to describe the manners of men, he writes:

> So that in the first place, I put for a general inclination of all mankind, a perpetual and restless desire of power after power, that ceaseth only in death. And the cause of this, is not always that a man hopes for a more intensive delight, than he has already attained to; or that he cannot be content with a moderate power: but because he cannot assure the power and means to live well, which he hath present, without the acquisition of more. (*E.W.* iii, pp. 85–6)

The prudent engine will always seek to secure itself, to the maximum extent possible, and so will desire 'power after power' as long as vital motion continues. This striving for power is not necessarily competitive—a solitary being, confronted by natural forces, would attempt to increase its power in relation to them, but could not be said to compete with them. But man is not a solitary being; he lives confronted by other men. Even this is not necessarily a competitive situation, for it is conceivable that the powers of different men should oppose only the powers of nature, and not each other. But once admit the possibility that one man's power opposes another, and the concern for 'power after power' becomes competitive. For if your power may be opposed to mine, may hinder my efforts to maintain my vital motion, then I must strive to secure myself by increasing my power in relation to yours. You, however, must equally strive to secure yourself by increasing your power in relation to mine. And so the race is on.

We can thus see why Hobbes supposes men to be necessarily competitors. Admit the *possibility* that men's powers be opposed, and you must admit the *necessity* of a struggle among them. This is the logic of all human conflict.

The detail of Hobbes's argument bears out this account; indeed, it presents the logic of conflict in a manner both lucid and elegant. Hobbes begins with the essential equality of all men—an equality which rests on the unpleasant fact that 'as to the strength of body, the weakest has strength enough to kill the strongest' (*E.W.* iii, p. 110). If survival is man's ultimate aim, then the ability of one man to kill another is the ultimate measure of man's equality. Hobbes then argues:

> From this equality of ability, ariseth equality of hope in the attaining of our ends. And therefore if any two men desire the same thing, which nevertheless they cannot both enjoy, they become enemies; and in the way to their end, which is principally their own conservation, and sometimes their delectation only, endeavour to destroy, or subdue one another. (*E.W.* iii, p. 111)

If men were manifestly unequal, the weaker would recognize their inability to compete in the power-struggle. They could maintain themselves, not by increasing their power in relation to others, for this would not be possible, but only by enlisting the power of some stronger person in their defence, in return for their homage to him.

Enmity arising out of competition leads to enmity arising out of diffidence. If you are my potential competitor, then it is in my interest to forestall you from the outset. As Hobbes says:

> And from this diffidence of one another, there is no way for any man to secure himself, so reasonable, as anticipation; that is, by force, or wiles, to master the persons of all men he can, so long, till he see no other power great enough to endanger him: and this is no more than his own conservation requireth, and is generally allowed. (Ibid.)

This is the crucial passage in Hobbes's account of conflict, for it explains the conversion of limited enmity into unlimited enmity. But there are yet other factors present in human conflict. Some men, Hobbes supposes, enjoy the exercise of their power in conquest without reference to their own security. Others must then increase their power to forestall such would-be aggressors. But Hobbes does not and need not suppose that all men are natural aggressors, exulting in conquest just because it affords them a display of their own power. Indeed, his explanation of enmity is sufficiently grounded in the claim that men aggress to better their own security; competition and enmity are merely intensified if some men are natural aggressors.

A further source of conflict is found in contempt.

> Again, men have no pleasure, but on the contrary a great deal of grief, in keeping company, where there is no power able to over-awe them all. For every man looketh that his companion should value him, at the same rate he sets upon himself: and upon all signs of contempt, or undervaluing, naturally endeavours, as far as he dares, (which amongst them that have no common power to keep them in quiet, is far enough to make them destroy each other), to extort a greater value from his contemners, by damage; and from others, by the example. (*E.W.* iii, p. 112)

The value of a man is the price others are prepared to pay for the use of his power. Contempt, then, indicates that others consider a man's power to be inferior to theirs. But no man can willingly admit such inferiority; he must assure himself of his ability to maintain himself, and convince others of this ability. Hence he must overcome his contemners, or suffer increased contempt, and increased inroads on his security.

> So that in the nature of man, we find three principal causes of quarrel. First, competition; secondly, diffidence; thirdly, glory.
> The first, maketh men invade for gain; the second, for safety; and the third, for reputation. (Ibid.)

Hobbes thus is led to his two famous metaphors: the comparison of life to a race, and the comparison of man's natural condition to a state of war. The race has

> no other *goal*, nor other *garland*, but being foremost, and in it: . . .
>> Continually to be out-gone, is *misery*.
>> Continually to out-go the next before, is *felicity*.
>> And to forsake the course, is to *die*. (*E.W.* iv, p. 53)

This metaphor is dangerously suggestive of the view that man's basic aim is to out-do his fellows, that he is innately and originally competitive. But as we have shown, competition is derivative; the innate and original concern with self-maintenance manifests itself as competition, but only because one man's power may always oppose another man's.

Similarly, the famous war 'of every man, against every man' is based, not on innate hostility, but on hostility derived from the ever-possible conflict between men's powers of self-maintenance. War is the consequence of natural insecurity, and the natural desire to preserve oneself. Hobbes's metaphor is grossly misunderstood if it is thought to show man's natural malevolence and evil.

But war, which arises out of the endeavour for self-preservation, is of course the greatest enemy of preservation. If all men seek to be foremost, most must fall behind. If all men fight, most must be overcome. The condition of war is the condition of 'continual fear, and danger of violent death; and the life of man, solitary, poor, nasty, brutish, and short' (*E.W.* iii, p. 113).

The natural condition of mankind is inherently unstable. The competitive search for increased security through increased power leads only to ever-increasing insecurity. Just as two nations seeking to strengthen themselves to prevent conflict with the other, find themselves locked in an arms race which tends to bring on that conflict, so men, seeking to strengthen themselves to prevent being overcome, find themselves locked in a race which ensures that most are overcome. The 'perpetual and restless desire of power after power' leads only to impotence. And impotence is death.

This is the predicament which underlies Hobbes's moral and political theory. The actions which men naturally and reasonably perform in order to secure their ends prove self-defeating. We cannot suppose that men in Hobbes's state of nature are irrational. They do not engage in the war of all against all merely in order to satisfy immediate passion, or even to secure short-term interests. In competing with their fellows they are seeking their over-all well-being. For the man who would opt out of the race, would put himself at the mercy of his fellows.[1]

It might be argued that, just as nations may ensure peace by never arming, so men might ensure their security by never seeking to forestall each other. But this proposal overlooks the ultimate cause of quarrel. Men become enemies because they desire the same commodities as needful to their preservation. If the state of nature were a state of plenty, then men might refrain from hostility. But given that a man, in order to survive, may need some object which is also needed by his fellows, then competition necessarily follows. And as Hobbes shows, diffidence follows competition.

Men acting on their own, however reasonable they may be, are doomed to the war of all against all. Hobbes therefore seeks to devise arrangements which will enable men to terminate this war. The well-grounded state is that condition of human affairs which enables men to overcome the obstacles to their security which make their natural condition one of permanent war.

But is any form of state possible, given the competitive character of Hobbesian men? Is it possible to overcome the diffidence which makes human intercourse an unending conflict? Hobbes was confronted by this objection, as his discussion in 'A Review, and Conclusion' to *Leviathan* shows. But it is not difficult to construct a plausible preliminary answer to it. Whether this answer will withstand a more searching examination of Hobbes's theories will emerge at the end of our study.

Were men, not self-maintaining engines, but other-destroying engines, or did their maintenance depend necessarily on the destruction of others, then war would be the inevitable condition of human existence. But if, by moral, political or

---

[1] We may compare men in the state of nature to firms in a state of perfect competition. Each firm seeks to maximize its share of the market; the result is to minimize the rate of profit, and so to minimize economic 'security'. But no firm can opt out of the race to maximize its share of the market; to do so is to put itself at the mercy of its competitors.

social arrangements, men can be enabled to maintain themselves without facing the competition of their fellows, then the basic cause of hostility among men will be removed.

Some men, Hobbes tells us, 'would be glad to be at ease within modest bounds' (*E.W.* iii, p. 112); give these men security, and they will willingly live at peace with their fellows. They are capable of society, although their natural condition prevents them from exercising this capacity.

But what of those who delight in conquest? Are they not innately hostile, and so do they not make war inevitable? Although their pleasure in exercising their own power is not a pleasure in what directly secures them, enhancing their vital motions, yet it is a pleasure in what better assures their security against possible though not present danger. The exercise of power in conquest reassures them of their ability to maintain themselves. Thus even these men, in a condition of security, may become tractable, provided that any attempt they make to exercise their power in conquest brings manifestly greater power against them. If their security is greater when living in modest bounds than in attempting to extend those bounds by conquest, they will no longer find pleasure in aggressive behaviour. And so they do not provide an insuperable problem to the peacemaker.

Some men naturally seek such power as is required to secure themselves, and find themselves forced by circumstances to seek ever more and more. Others naturally seek power without limit, finding such power a reassurance against possible hazards. These men may behave irrationally, in that the diffidence with which they view their fellows may not be grounded in a considered opinion of those fellows as possible competitors. But even these men do not act in a manner deliberately and directly inimical to their own security. Thus, when security is assured to those who remain within modest bounds, but denied to those who seek 'power after power', even they will cease to seek power at the expense of others, and so will adjust themselves to the conditions of peace.

Men are like those beasts which are naturally wild, but

capable of being tamed. Left to themselves their condition is one of endless conflict. But trained, and reined, they need not be confined to separate cages in a zoo; they can live in each other's company. Hobbes's problem is to devise effective training, and effective reins.

5. HUMAN NATURE AND POLITICAL SYSTEMS

Early in our inquiry into Hobbes's account of human nature we raised the question of its bearing on his political theory. Hobbes supposes that by examining the parts of the Commonwealth—the individual men who are both its matter and its makers—he can derive the principles which determine the proper composition of the whole. He can advise men what to do in order to set up a well-grounded state. But this supposition, it is objected, confuses a descriptive or factual account with a prescriptive or normative proposal.

We saw that this objection does not deny the relevance of the descriptive account of human nature to the prescriptive proposal. Rather it denies the claim that the proposal can be deduced, or formally inferred, from the account. Before attempting to assess the objection in the light of our account of man, let us put it as forcefully as possible, in the form of a dilemma.

Hobbes's conclusions must meet two conditions. First, they must be formally inferred from his account of human nature— that is, primarily from his account of human motivation. From statements about motives Hobbes derives statements about behaviour.

Secondly, they must lay down a pattern of behaviour which man can follow, but need not. For clearly Hobbes believes that men are not behaving as they should, but can do so. His conclusions are intended to provide advice to men to *change* their behaviour so that they may succeed in constructing a well-grounded state.

The first condition requires that Hobbes's conclusions be true descriptions of how men do behave, provided his premises

are true and his inferences sound. The second condition requires that these same conclusions not be true descriptions
of how men do behave, but only descriptions of how men may
behave. These requirements can not both be satisfied. Hobbes
can provide either an account of behaviour which has no
necessary role in political instruction, or a piece of advice
which is not deduced from his account of human nature.

This objection is not without some foundation. Hobbes
undoubtedly did not recognize the difference between physics
and politics, between the connection of gears, wheels, and
levers with a mechanical engine, and the connection of men
with a Commonwealth. But this is a merely negative consideration. We must ascertain what can be derived from Hobbes's
account of man, so that we may determine the extent to which
his procedure is inadequate.

If we accept Hobbes's view that man is a self-maintaining engine, then we can establish the basic nature of human
motivation. Men want, and necessarily want, to preserve
themselves. Therefore, whatever can be shown to be a condition of human preservation, is thereby shown to be a means
to man's end. From premises of the form 'X is a necessary
means to self-preservation', Hobbes can derive conclusions of
the form 'a man must do X to secure what he wants'.

If we accept also Hobbes's contention that 'those actions
are most reasonable, that conduce most to their ends' (*E.W.*
iii, p. 133), we can then derive from 'a man must do X to
secure what he wants', the further conclusion 'a man, if
rational, will do X'. This statement is an ideal description, a
description of how a fully rational man will necessarily behave.
It is not a description of actual human behaviour because men
are not fully rational, mistaking both their ends and the means
to them.

From the claim 'a man, if rational, will do X', it seems
possible to derive the imperative of advice 'Do X!' For any
man has sufficient reason to do whatever he would do, were he
fully rational, and what he has sufficient reason to do is what
he is best advised to do.

The connection between 'a man, if rational, will do X' and 'Do X!' is not logical entailment. A statement can not entail an imperative, for much the same reason that a judgement that one should perform an action cannot entail a decision to perform the action. But we may say that just as the judgement fully justifies the decision, in that a man who judges that he should perform some action behaves irrationally if he does not decide to perform it, so the statement 'a man, if rational, will do X' fully justifies the imperative 'Do X!', in that a man who accepts the statement is behaving irrationally if he rejects the imperative.

Thus, in so far as Hobbes's political doctrine tells men what they must do in order to preserve themselves, Hobbes can claim that acceptance of its imperatives is rationally required, if one accepts the psychological doctrine that self-preservation is man's basic end.

But Hobbes's doctrines are not couched merely in the language of advice. He talks of the *right* that each man has to act in certain ways, of *laws* of nature, of the *obligation* to obey the civil sovereign. These terms, 'right', 'law', and 'obligation', which provide Hobbes with his basic moral vocabulary, are used to develop his political doctrine. Grant for the moment that his psychology does sufficiently justify advice; how can it justify moral conclusions?

This question we must postpone. Until we examine Hobbes's moral and political system we cannot determine whether his psychology can adequately justify it. But to anticipate the principal consideration we shall introduce, Hobbes's moral terms are themselves introduced by way of the phrase 'right reason', with which we are already familiar. Hobbes's account of man as rational provides the key link between his psychological doctrines and his moral and political conclusions.

Our concern so far has been to answer the charge that Hobbes commits a logical fallacy in deriving normative conclusions from factual premisses. In characterizing man as rational, Hobbes provides for the derivation of judgements which describe how an ideal man would behave, but prescribe how actual men should behave. And this seems a quite legitimate procedure.

But it would be a mistake to conclude that Hobbes does not confuse factual and normative issues. The confusion occurs, however, not in the derivation of moral and political conclusions from psychological premisses, but in the premisses themselves. And here two quite distinct objections present themselves.

The first is to Hobbes's claim that men are necessarily self-maintaining engines, that self-preservation is a necessary and basic motive of human action.

We are frequently in a position to establish singular motivational statements, in which we explain a particular action of a particular person in terms of a particular motive. Self-preservation enters not uncommonly into such statements. We are also frequently in a position to establish general motivational statements, in which we explain many actions of many persons in terms of a certain type of motive. Once again, self-preservation may figure in such statements.

General motivational statements enable us to predict that, in a given situation, a person will very likely act in a certain way because he will very likely be motivated by such-and-such a factor. But we should not predict that a normal person in normal physiological and psychological condition—not drugged nor hypnotized, for example—*must* act in a certain way because he *must* be motivated by such-and-such a factor. Yet this is what Hobbes endeavours to do; his theory requires that a normal person must perform those actions which he believes requisite to his preservation because he must be motivated by a concern for maintaining himself.

Empirically this is an untenable position. Considering suicide, Hobbes is forced to say:

But I conceive not how any man can bear . . . so much malice towards himself, as to hurt himself voluntarily, much less to kill himself. For naturally and necessarily the intention of every man aimeth at somewhat which is good to himself, and tendeth to his preservation. And therefore, methinks, if he kill himself, it is to be presumed that he is not *compos mentis*, but by some inward torment or apprehension of somewhat worse than death, distracted. (*E.W.* vi, p. 88)

But what can be worse than death for a being whose sole aim is to live? Hobbes must suppose that the functioning of the self-maintaining engine is in some manner impaired.

Furthermore, and even less consistently with his theory of human motivation, Hobbes admits that death is not always the greatest evil to a man who is *compos mentis*. A man can never be obliged to kill his parent since 'a son will rather die than live infamous and hated of all the world' (*E.W.* ii, p. 83).

We may suppose that Hobbes has simply failed to give a sufficiently complete account of man's basic motivations. We may hold that we do not predict that a person must be motivated to act in a particular way, because human motivation is too complex for us to be certain what factors will be operative in any situation. But it seems to me that the objection we are now considering to Hobbes's theory of motivation is not based on the inadequacy of the theory. Any theory of motivation will, I suggest, show us only what a man in normal circumstances *may* do, not what he *must* do. But this suggestion is not one which can be developed here.

If this first objection be accepted, then we must modify our account of what Hobbes can derive from his theory of human nature, for we must modify that theory itself. We must say, not that man is a self-maintaining engine, but that he is an engine which tends to seek to maintain itself. In other words, we may say that self-preservation is a very important human motive, but not the necessary basic motive. And so from the premiss 'X is a necessary means to self-preservation' we can derive only the conclusion 'a man must do X to secure what, very probably, he wants'. Hobbes's psychological theory will still support his normative conclusions, but the support will be substantially weakened.

The second objection is to Hobbes's account of human rationality. We need not concern ourselves here with the adequacy of Hobbes's mechanistic account of reasoning. Rather, we must consider the status of the contention, stated earlier, that 'those actions are most reasonable, that conduce most to their ends'.

Hobbes requires this in order to pass from statements about what is necessary to secure a man's end to statements about what the man, if rational, will do. But can we accept this further premiss as a factual or descriptive statement about human nature?

We may agree with Hobbes. But our agreement is not simply an assent to the facts. Moral action is often thought to conduce, not to a man's ends, but to the ends of other men, or of society, or of God; yet moral action is not thought to be against reason. Now it may be that such a view of morality is mistaken, but it cannot be ruled out of court simply by an alleged factual claim that reasonable actions are those which conduce to a man's ends. This claim embodies a fundamental normative contention.

Once again, then, we must modify our account of what Hobbes can derive from his theory of human nature. I suggest that we modify Hobbes's premiss about rationality to read 'actions which conduce to the agent's ends are in this respect reasonable', leaving open that other factors as well may bear on the reasonableness of action, and may in some cases override a man's own ends. Even this involves a normative claim, but hardly a disputable one.

Thus from the premiss 'X is a necessary means to self-preservation' we derive 'a man must do X to secure what, very probably, he wants'. And from this we derive 'a man very probably has some reason to do X'. This will provide a reason, though by no means always a sufficient reason, for the imperative 'Do X!'

Although this modification shows that Hobbes makes much too strong a claim, in purporting to demonstrate a moral and political system on the basis of his account of human nature, it does not at all suggest that Hobbes's psychology is not essential to his morals and politics. We are not yet in a position to consider this much more radical objection to what Hobbes takes to be his method.

The objections we have considered in this section are not, of course, the only or the most important criticisms one might make of Hobbes's view of human nature. Our interest is only in objections *in principle*—objections to the structure of Hobbes's argument, and not to its actual content. In so far as Hobbes's view of human nature is mistaken, we may expect his moral and political doctrines to be in some ways inadequate. But it by no means follows that his doctrines lack interest. For as we shall see, the very uncompromising view of man with which Hobbes begins imparts to his moral and political theory a clarity and rigour which may help us in the understanding of other, ultimately more realistic, theories.

II

MORAL THEORY

HOBBES writes on morals as a moralist. His primary aim is to demonstrate what men ought, and what they ought not, to do. In pursuing this aim he introduces, and explains, certain moral concepts, of which the most important are *right of nature*, *law of nature*, *obligation*, and *justice*. But his interest is in using these concepts in moral conclusions, not in explicating them.

When the contemporary moral philosopher turns to Hobbes, however, his concern is with the concepts themselves, rather than the conclusions in which they appear. He wants to analyse moral terms, not draw moral conclusions. This shift of interest is entirely legitimate, but it is not always sufficiently noticed.

If asked for his moral theory, or moral doctrine, or moral philosophy, Hobbes might answer briefly in this way:

> (M)oral philosophy is nothing else but the science of what is *good*, and *evil*, in the conversation, and society of mankind. . . . (*J*)*ustice*, *gratitude*, *modesty*, *equity*, *mercy*, and the rest of the laws of nature are good; that is to say; *moral virtues*; and their contrary *vices*, evil. Now the science of virtue and vice, is moral philosophy; and therefore the true doctrine of the laws of nature, is the true moral philosophy. (*E.W.* iii, p. 146)

But if we ask one of the modern commentators for Hobbes's moral theory, we may receive an answer beginning: 'Hobbes's ethical doctrine proper, disengaged from an egoistic psychology with which it has no logically necessary connection, is a very strict deontology . . .'[1] And predictably, this answer has not been sufficiently distinguished from the type of answer Hobbes would give. Hence we find another commentator

---

[1] The commentator is A. E. Taylor, and the passage from 'The Ethical Doctrine of Hobbes' appears, reprinted, in *Hobbes Studies*, p. 37.

telling us, 'But few have ignored Taylor's principal contention that Hobbes's ethical theory, considered on its merits
and as distinct from the psychology, is in some sense a deontology and not an explication of moral and political obligation
in terms of prudence and self-interest.'[1]

Now it is one of the principal theses of this study that
statements such as that just quoted misfire in an important
way. Of course Hobbes is trying to convince us that we have
certain moral and political obligations, and the arguments he
uses are based on considerations of prudence and self-interest.
This has been denied by some commentators, but the denial is
absurd on the face of it, and absurd on reflection. But it does
not follow that the moral—or *practical*—concepts employed
by Hobbes are therefore to be *defined* in terms of prudence or
self-interest. As we shall show, in the proper sense of definition, they are not to be taken as prudential concepts.

There are two opposed lines of argument which have been
pursued by Hobbes's scholars, and which are equally mistaken.[2] The first is to claim, correctly, that Hobbes's *conclusions* have a prudential basis, and to infer, incorrectly, that
Hobbes's *concepts* are prudential. The second is to claim,
correctly, that Hobbes's concepts are not prudential, and to
infer, incorrectly, that Hobbes's conclusions do not have a
prudential basis.

The plausibility of these mistakes has been increased by
two factors. The first is the supposition, usually tacit, that if
Hobbes's concepts are not prudential, and not merely legal,
then they must be moral in a sense of 'moral' correlative but
opposed to that of 'prudential'. This is wrong; as we shall
show, Hobbes's concepts are *practical*, moral in so far as
'moral' means 'practical', 'concerning what to do', but not
in so far as 'moral' means 'opposed or superior to prudential'.

[1] Stuart M. Brown, Jr., 'The Taylor Thesis, Introductory Note', in *Hobbes
Studies*, p. 31.

[2] As recent examples of the first, Stuart M. Brown, Jr., 'The Taylor Thesis:
Some Objections', and John Plamenatz, 'Mr. Warrender's Hobbes' in *Hobbes
Studies*. As examples of the second, A. E. Taylor, op. cit. and Howard Warrender,
op. cit.

The second is the supposition that Hobbes's concept of obligation must be defined in terms of his concept of the law of nature. This we shall show to be mistaken—and badly mistaken, in that it rests on a misunderstanding of our ordinary concept of obligation.

So much by way of promise of what is to come. Our procedure will be as follows. First, we shall extract, from Hobbes's account of his moral concepts, what I term the *formal definitions* of these concepts. These definitions will show the logical independence of Hobbes's moral concepts from his psychology, and will show also that these concepts are practical, not merely prudential in force.

Second, we shall put flesh on the bones of these formal definitions by considering what I term the *material definitions* of Hobbes's basic moral concepts. These definitions do depend on Hobbes's psychological theory. In terms of these definitions we shall be able to outline, as far as is relevant for our purposes, Hobbes's moral system—his doctrine of what we ought, and ought not, to do.

Third, we shall examine a number of problems about Hobbes's concepts, understood both formally and materially. Here we shall make good the claim that laws of nature and obligation are formally, although not materially, independent in Hobbes's theory. But we shall also note a grave problem which arises for Hobbes, in employing his concept of obligation.

Finally, we shall consider whether Hobbes can be said to have a moral theory. Here again, our distinction between formal and material definitions will prove necessary to distinguish issues involved and enable us to bring this part of our study to a clear conclusion.

I. THE MORAL CONCEPTS: FORMAL DEFINITION

A. *The Right of Nature*

The right of nature is introduced at the beginning of the fourteenth chapter of *Leviathan*. No prior explication of the concept is given; it is simply defined as follows:

The RIGHT OF NATURE, which writers commonly call *jus naturale*, is the liberty each man hath, to use his own power, as he will himself, for the preservation of his own nature; that is to say, of his own life; and consequently, of doing any thing, which in his own judgment, and reason, he shall conceive to be the aptest means thereunto. (*E.W.* iii, p. 116)

Hitherto Hobbes has considered man as an animal, or an engine. He has established man's basic motivation, and derived its behavioural consequences, which culminate in the war of all against all. But now, without warning, he considers man as a moral being, possessed of a certain right.

The passage we have quoted contains both the formal and material definition of this right. In order to separate out the purely formal part, we must first explain certain features of the definition which may not be immediately apparent, and then provide the prior explication underlying the introduction of the concept which, as we shall show, Hobbes has simply omitted from *Leviathan*.

One feature of Hobbes's definition we shall omit from discussion here. The word 'liberty', as scholars have recognized,[1] is not defined and used by Hobbes in a clear and consistent manner. We shall turn to the problems it raises in the third part of this chapter; here it is sufficient to take the word in its obvious sense, and assume that it conveys permission—what a man *may* do.

What is in accord with the right of nature, then, is what one may do, what it is *all right* for one to do. It is not, however, what one has *a right to*, a claim which must, or ought to, be recognized by others. The right of nature entails no correlative duties.

Most philosophers and persons in general who talk about natural rights or human rights do intend such an entailment. An appeal to the natural right of the child to live, in arguments about abortion, is relevant because if this right is granted it entails a duty on the part of others—doctors, for

---

[1] For example, J. R. Pennock, 'Hobbes's Confusing "Clarity"—The Case of "Liberty"', and A. G. Wernham, 'Liberty and Obligation in Hobbes', both in *Hobbes Studies*.

example—to do what is necessary for the child to live, and
not to do what is incompatible with the child's living. If life,
liberty, and property are, as some have thought, natural rights,
then at the least no one ought to deprive another of his life
or liberty or property.

But Hobbes's conception of right has no such entailment.
Hobbes claims that I have the right to do whatever I conceive
necessary for my own preservation—in this sense, certainly,
I have the natural right to live. But no one therefore has
the duty to allow me to do what I consider necessary to my
preservation. You, having the right to do what you consider
necessary to your preservation, may judge my demise advantageous, and so may with right kill me. You deprive me of my
life, and prevent my exercise of my right of nature, but you
do not violate my right of nature.

Thus we may say that, for Hobbes, the following simple
equivalence holds:

'A has the right to do X' = 'A may do X'.

To include a reference to the right *of nature*, we may add to
the equivalence in this way:

'A has the natural right to do X' = 'A may initially do X'.

The introduction of 'initially' is intended to signal the fact
that for Hobbes the right of nature is the right possessed by
each man prior to any events which serve to limit that right.
The character of these events will demand explanation in
due course.

In our statements of equivalence we have suppressed all
reference to preservation. Such reference belongs properly to
the material definition of the right of nature. To insert it, so
that the equivalence read:

'A has the natural right to do X' = 'A may initially do X
    for the preservation of his own nature'

would make no *material* difference, but it would have the
formal consequence of committing Hobbes to denying that, if

there were an initially permissible action directed to some end other than preservation, one would have the natural right to do it. And it is surely unlikely that anyone, Hobbes included, would want to deny this.

To complete our analysis of the formal meaning of 'right of nature', we must turn to a discussion which appears in Hobbes's earlier writings, but not in *Leviathan*. Here, right of nature is introduced in connection with reason—a connection which is implicit in the argument of *Leviathan*, although not directly stated. The key passages are these:

> . . . it is not against reason, that a man doth all he can to preserve his own body and limbs both from death and pain. And that which is not against reason, men call *right*, or *jus*, or *blameless liberty* of using our own natural power and ability. It is therefore a right of nature, that every man may preserve his own life and limbs, with all the power he hath. (*E.W.* iv, p. 83)

What is not against reason is called right. This connection is reinforced in *De Cive* by using the intermediate concept of *right reason*:

> For every man is desirous of what is good for him, and shuns what is evil, but chiefly the chiefest of natural evils, which is death; and this he doth by a certain impulsion of nature, no less than that whereby a stone moves downward. It is therefore neither absurd nor reprehensible, neither against the dictates of true reason, for a man to use all his endeavours to preserve and defend his body and the members thereof from death and sorrows. But that which is not contrary to right reason, that all men account to be done justly, and with right. Neither by the word *right* is anything else signified, than that liberty which every man hath to make use of his natural faculties according to right reason. (*E.W.* ii, pp. 8–9)

Against right reason Hobbes admits no appeal. Whatever a man does in accordance with reason, he does permissibly and blamelessly. Hence we may adopt the equivalence:

'A may do X' = 'A doing X is in accordance with (right) reason'.

And combining this with our first equivalence, or being

guided directly by the last sentence of the passage from *De Cive*, we have:

'A has the right to do X' = 'A doing X is in accordance with (right) reason',

and:

'A has the natural right to do X' = 'A doing X is initially in accordance with (right) reason'.

This last equivalence is our contextual formal definition of Hobbes's concept of the right of nature. It sets Hobbes's basic moral concept in its proper context—rationality in the sphere of action. But it does not import into the concept the content of Hobbes's psychological theories. And so its acceptability is in no way dependent on the belief that action is necessarily egoistic. Indeed, the definition itself seems quite unobjectionable; the difficulties clearly arise in its application.

There is, however, an objection to it, in terms of Hobbes's text. For he says 'RIGHT, consisteth in liberty to do, or to forbear' (*E.W.* iii, p. 117). To make our definition conform to this passage, we should have to recast it as:

'A has the natural right to do X' = 'A doing or not doing X is initially in accordance with (right) reason'.

Only by adding 'or not doing' can we bring in the force of 'or to forbear'.

But this alteration would give rise to absurdities. For surely an action may accord with right reason, without its omission also according with right reason. On this definition we should be forced to say either that one did not have the natural right to perform such an action, or that every action one might have reason to perform was such that one might also have reason to omit it.

Hobbes frequently insists on the inalienable natural right to defend oneself against immediate threats to life. On this definition Hobbes would mean that defending or not defending oneself against such threats is in accordance with right reason. But as we shall see, not to defend oneself is neither rational nor psychologically possible. We must reject the proposed alteration, then, as inconsistent with Hobbes's basic position.[1]

Before proceeding, we should consider one question which our account clearly invites. Why does Hobbes omit from *Leviathan*, his most developed and mature political work, the arguments which we have used to construct our definition? Is there not the most uncomfortable possibility that Hobbes omitted these arguments because he no longer accepted them?

I think not. The account in *Leviathan*, as we have seen, leaves the concept of the right of nature entirely without prior explication. And surely some indication is needed of how it is to be introduced into an argument which, to this point, has treated man only as a self-maintaining engine. Nothing in the arguments present in *Leviathan* is incompatible with the supposition that the right of nature is based on reason, and so we are surely entitled to turn to the earlier works for our definition.

We can, indeed, give some account of the omission of the derivation of right of nature from *Leviathan*. In *De Corpore Politico* and *De Cive*, Hobbes introduces the right of nature prior to his discussion of the state of nature, and then shows

---

[1] A. G. Wernham, op. cit., pp. 129–30, attempts to analyse the right of nature in a way compatible with Hobbes's claim that 'RIGHT, consisteth in liberty to do, or to forbear'. He says, 'Thus when Hobbes says that an agent did x by natural right we must take him to mean that it was permissible for that agent to do x or forbear x according as he should think necessary for his preservation, and that he thought it necessary for his preservation to do x.' Distinguishing formal from material elements, we have:

'A does X by the right of nature' = 'It is permissible for A to do X or forbear X as he thinks reasonable, and he thinks it reasonable to do X'.

But then to define the right of nature itself, Wernham must surely be committed to:

'A has the right to do X' = 'It is permissible for A to do X or forbear X as he thinks reasonable'.

And from this it follows that if A has the right to defend himself, he has also the *right* not to defend himself, though he will not think this reasonable. But to admit such a right, or to divorce reason from right in this way, is completely contrary to Hobbes's statements.

how the right of nature, conceived as an initial unlimited right, leads to the war of all against all. In the state of nature reason is the slave of the passions; hence to refer to a right to do what one naturally endeavours to do is otiose.

In *Leviathan*, then, Hobbes introduces human rationality, as manifest in action, only when he reaches that stage in his argument at which reason and passion diverge—only when he begins to consider how man can rationally escape from the impasse created by his passionate nature. This order of exposition is clearly superior to that in the earlier writings. But in reconstructing his argument in this way, Hobbes divorces discussion of right reason from the introduction of the right of nature, for the former is part of his account of individual human nature. And he omits to add connecting links, which would make explicit the connection of right reason and the right of nature. In this way his improved order of exposition has understandably misled some of his readers.

B. *Law of Nature*

The concept of a law of nature is perhaps the most difficult of Hobbes's fundamental moral terms to explicate. Our treatment in this section will be in some respects dogmatic; we shall postpone until the third part of this chapter a full discussion of some of the major problems involved. A justification of the formal definition we shall give depends upon determining the actual use to which Hobbes puts the concept, and this can be found only in the conclusions which Hobbes reaches about the content of the laws of nature and their practical relevance. These consequences depend on the material definition of the concept, which we shall state in the next part of the chapter.

In *Leviathan*, Hobbes presents us with this definition:

A LAW OF NATURE, *lex naturalis*, is a precept or general rule, found out by reason, by which a man is forbidden to do that, which is destructive of his life, or taketh away the means of preserving the same; and to omit that, by which he thinketh it may be best preserved. (*E.W.* iii, pp. 116–17)

To bring out the formal aspects of this account, we must turn again to Hobbes's earlier writings, where the connection with reason is explained:

> But since all do grant, that is done by *right*, which is not done against reason, we ought to judge those actions only *wrong*, which are repugnant to right reason, that is, which contradict some certain truth collected by right reasoning from true principles. But that which is done *wrong*, we say it is done against some law. Therefore *true reason* is a certain *law*; which, since it is no less a part of human nature, than any other faculty or affection of the mind, is also termed natural. Therefore the *law of nature*, that I may define it, is the dictate of right reason, conversant about those things which are either to be done or omitted for the constant preservation of life and members, as much as in us lies. (*E.W.* ii, pp. 15–16)

The laws of nature are rational precepts, laying down what reason *requires*, rather than merely *permits*. Or put another way, what is contrary to the laws of nature is what is contrary to reason; we have the equivalence:

'X is contrary to the laws of nature' = 'Doing X is contrary to (right) reason'.

To speak of these precepts as laws would seem, from the passage we have quoted from *De Cive*, to be only to say that action against them would be wrong. Since 'wrong' means only 'repugnant to right reason', we have the simple equivalence:

'law of nature' = 'precept laying down the requirements of (right) reason'.

And this is our formal definition of a law of nature.

But there are two immediate objections to this definition. The first is that, so defined, the laws of nature are not properly laws. Hobbes insists that 'law, properly, is the word of him, that by right hath command over others' (*E.W.* iii, p. 147). Hence our definition omits what is essential to the understanding of the laws of nature as laws.

The second objection is that Hobbes presents us with an alternative account of the laws of nature, saying that they are 'delivered in the word of God, that by right commandeth all things' (ibid.). This account suggests an alternative equivalence:

'law of nature' = 'command delivered in the word of God'.

And this alternative will meet the first objection; if God commands by right, then his commands are properly laws.

A full assessment of these objections will be postponed. But it is possible to present an initial defence of our definition by considering the full passage from which we have quoted the two excerpts expressing the objections. It reads:

These dictates of reason, men used to call by the name of laws, but improperly: for they are but conclusions, or theorems concerning what conduceth to the conservation and defence of themselves; whereas law, properly, is the word of him, that by right hath command over others. But yet if we consider the same theorems, as delivered in the word of God, that by right commandeth all things; then are they properly called laws. (Ibid.)

This passage suggests that, in accordance with our formal definition, the laws of nature, as introduced by Hobbes, are not laws. The second definition is based on an alternative way of regarding these precepts, and from this point of view the precepts are laws. But if our definition is to be based on Hobbes's initial account, then it would be mistaken for it to be a definition of *law*.

The question then is which of these two ways of looking at the laws of nature is basic to Hobbes's moral theory. This question can be settled only by considering the content of that theory, but we shall find that the first way proves basic. And this will justify our formal definition.

We may, however, present three considerations here in support of our claim that Hobbes's first definition is basic. The first is taken from his order of exposition. The first definition serves to introduce the concept; Hobbes then proceeds to develop his doctrine of the laws of nature, stating their content fully, on the basis of this introduction. Only when he has completed his exposition does he introduce the second definition. It would be odd, to say the least, to suppose then that the second definition was in fact essential to the theory.

In the second place, Hobbes equates his order of exposition with the temporal order in which man has come to understand the laws of nature. In his *Answer to Bramhall*, in which he defends *Leviathan* against the Bishop's objections, he says: 'I thought it fittest in the last place, once for all, to say they were the laws of God, then when they were delivered in the word of God; but before, being not known by men for any thing but their own natural reason, they were but theorems, tending to peace' (*E.W.* iv, pp. 284–5). Since the whole concern of Hobbes's moral and political philosophy is to show men the way out of the war of all against all and into the condition of peace, it is surely evident from this passage that Hobbes takes the first definition to be essential, and the second definition to be indeed secondary, and not required for his purpose.[1]

And in the third place, if Hobbes thought that the second definition was basic, he would surely have given careful attention to showing that the laws of nature are indeed the commands of God. Now Hobbes does take pains to show that the laws of nature are contained in Scripture, so that he may meet religious objections to his theory. But this shows only that the laws of nature are binding on Christians as scriptural laws, not that they are natural laws. In fact, as we shall see in Chapter V, Hobbes has no arguments in support of the claim that the laws of nature are naturally delivered as commands of God.

We may suggest the following conclusion. If our formal definition of a law of nature is mistaken, then Hobbes is mistaken about what is required of the laws of nature if they are to play their role in his moral and political system. This is,

---

[1] This passage is sufficient to refute the interpretation of Hobbes given by F. C. Hood, *The Divine Politics of Thomas Hobbes*, Oxford, 1964. Hood thinks that Hobbes is writing only for Christians. On the contrary, Hobbes is writing for rational men. Christians may, of course, be rational, but there seem to be rational men who are not Christians.

of course, possible. But we shall show that Hobbes is not mistaken.

Before leaving our formal account of a law of nature, we should consider briefly the formal relationship between the concepts of the right of nature and the laws of nature. And here a difficulty intrudes.

Given our formal definitions, it would seem that the laws of nature are precepts instructing us in the exercise of the right of nature. That is, we may do whatever accords with reason; the laws of nature advise us what not only accords with reason, but is required by it. Hence acting on the laws of nature is acting rightly, exercising the right of nature, but in those special circumstances in which acting otherwise would be acting wrongly.

Furthermore, the laws of nature would not themselves impose limitations on the right of nature. In telling us what is rationally required, they would serve the useful role of enabling us more effectively to act in accordance with reason, or to exercise the right of nature, but they would not affect that right in principle.

But unfortunately, what Hobbes says is rather different.

> For though they that speak of this subject, use to confound *jus*, and *lex*, *right* and *law*: yet they ought to be distinguished; because RIGHT, consisteth in liberty to do, or to forbear; whereas LAW, determineth, and bindeth to one of them: so that law, and right, differ as much, as obligation, and liberty; which in one and the same matter are inconsistent. (*E.W.* iii, p. 117)

However, we have already seen that 'liberty to do, or to forbear', cannot be imported into Hobbes's definition of right without creating absurdities. And it is this phrase in the passage in question which enables Hobbes to oppose right and law in the way in which he does.

If then we may ignore this way of explaining the difference between right and law, as involving an inconsistency with more basic features of Hobbes's moral theory, we may accept the relationship between the laws of nature and the right of nature implied by our formal definitions. But, although we shall maintain that it is strictly accurate to say that the laws of nature do not themselves impose limitations on the right of nature, we shall see that this claim may easily be misunderstood, and as misunderstood it becomes clearly false.

C. *Obligation*

In order to provide a formal definition of obligation, we do not have to abstract from a more comprehensive account of the concept actually found in Hobbes's argument. Instead, we have to construct our definition from a rather meagre, but purely formal, base. The primary sources are these:

(i) 'obligation, and liberty . . . in one and the same matter are inconsistent'. (*E.W.* iii, p. 117)

(ii) 'To *lay down* a man's *right* to any thing, is to *divest* himself of the *liberty*, of hindering another of the benefit of his own right to the same.' (*E.W.* iii, p. 118)

(iii) 'And when a man hath . . . abandoned, or granted away his right; then is he said to be OBLIGED, or BOUND, not to hinder those, to whom such right is granted, or abandoned, from the benefit of it: and that he *ought*, and it is his DUTY, not to make void that voluntary act of his own.' (*E.W.* iii, p. 119)

(iv) 'there being no obligation on any man, which ariseth not from some act of his own; for all men equally, are by nature free'. (*E.W.* iii, p. 203)

These passages suggest the following doctrine. To grant away one's right to perform some action, or to possess some object, is to undertake an obligation not to perform that action, or to possess that object. When, and only when, one has granted away one's right in some matter, one has an obligation in that matter.

All obligations, then, are self-imposed. There are no natural obligations, co-ordinate with natural rights. There are no obligations outside the area of the right of nature; if one has an obligation not to do some action, then previously one had a natural right to do that action.

Why obligations are imposed—why men restrict their rights—is a question falling within the material part of

Hobbes's theory. The formal definition of obligation is logically independent of why, or even whether, men do restrict their rights. It would seem to be simply:

'A has an obligation not to do X' = 'A has laid down the natural right to do X'.[1]

That this simple and elegant account of obligation has not been duly appreciated by the majority of Hobbes's interpreters is, I think, astonishing. For with this definition we may explain at once how obligation can be introduced into a situation in which no prior obligations exist. At one blow Hobbes can, and does, cut away the arguments of those who say that political obligation—which is what Hobbes wishes primarily to explain—must be derived from some prior obligation to obey the will of God, or to follow the laws of nature, or what have you. Hobbes shows, on the contrary, that political obligation—and with it all obligation—is a human creation, the result of certain actions which men can and do perform.

The principal device by which men oblige themselves is *covenant*. A covenant is that species of contract, or 'mutual transferring of right' (*E.W.* iii, p. 120), in which at least one of the parties 'is to perform in time to come' (*E.W.* iii, p. 121). 'Mutual transferring of right' must be understood in a rather special sense, since Hobbesian rights—permissions or liberties—cannot literally be transferred. What Hobbes intends is that each party to the covenant agrees not to oppose the exercise of some right by the other, and this is achieved by laying down his own corresponding right.

Covenants, then, oblige necessarily; to covenant is to lay down a certain right, which is to assume an obligation. But this point is just misunderstood. J. Roland Pennock, for example, writes:

Stuart Brown, Jr. contends that it is absurd to argue that grounding obligation on covenants implies a prior principle regarding the validity of covenants, because the very notion of covenants *implies* obligation to perform. But surely it must imply an obligation independent of, and

[1] Cf. A. G. Wernham, op. cit., p. 132.

prior to the covenants; hence this leaves unanswered the question of how it is possible to have covenants in a situation (state of nature) in which no obligations exist.[1]

It is unfortunately clear that Pennock has simply missed the entire point of Hobbes's account. The claim that covenants imply obligation to perform is just the claim that covenants do not require a prior and independent obligation in order to oblige. And it is this claim which Hobbes needs, as we shall see, to develop his moral system.

It may be possible to clarify the misunderstanding created by Hobbes's account by taking as an example our ordinary obligation of promise-keeping. To show that I have an obligation to do some act, X, which I have promised to do, I might reason as follows:

> Factual premiss: I promised to do X.
> Partial definition: To promise is to put oneself under obligation.
> First conclusion: Therefore I put myself under obligation to do X.
> Further factual premiss: I have not discharged, nor been released from, the obligation to do X.
> Second conclusion: Therefore, I am under obligation to do X.

As this reasoning explicitly shows, promises oblige for us in just the way covenants do for Hobbes. When we promise, we put ourselves under obligation. Prior to making the promise we were under no obligation in the matter in question—and may, for all that it matters, have been under no obligations whatsoever. But after promising, we are under obligation.

But, it may be asked, what of the principle of promise-keeping? Are we not under a general obligation to keep our promises, and is this not prior to the specific acts of promising which we perform? Not at all. The principle of promise-keeping is simply a reformulation of the definition above; the so-called general obligation to keep our promises is no more

---

[1] J. R. Pennock, op. cit., p. 112 n.

than the sum of the specific obligations we assume in promising. We do not reason in this way:

Factual premiss: I promised to do X.
Moral principle: Promise-keeping is obligatory.
Conclusions, etc., as before.

For this argument would simply invite the question 'Why is
promise-keeping obligatory?' And the answer to this must be
'Because promising is putting oneself under obligation'. The
principle proves to be an obscure stand-in for the partial
definition.[1]

We should note that prior to and independent of our acts
of promising, we must have the *capacity* to oblige ourselves.
This capacity is not itself an obligation. The claim that men
can oblige themselves by promising is not itself the ground
of any obligation whatsoever; only an act of promising will
produce an actual obligation.

Similarly, Hobbes must suppose that men have the capacity
to oblige themselves, that is, to grant away certain of their
rights. We shall see later that this supposition creates real
problems for Hobbes, given his view of the nature of man.
But it does not reveal any inadequacy in his actual definition.

There is, however, one possible inadequacy in the definition,
which we must briefly consider. Suppose a man renounces a
right, but does so contrary to right reason. Does he thereby
put himself under obligation?

Hobbes never considers this question, for his concern is
with situations in which men have reason to renounce their
rights. But if we were to suppose that he would give a negative
answer, insisting that a wrongful laying down of right was of
no effect, then our contextual definition should read:

'A has an obligation not to do X' = 'A has laid down the
natural right to do X, in accord with right reason'.

---

[1] Of course someone will suggest the answer 'Because promising is justified
on utilitarian grounds', or something of the sort. But this is to confuse the
rationale for the institution of promising, with the character of the institution.
Why do we have this way of obliging ourselves? For utilitarian reasons, perhaps.
But we can not then use these same reasons to justify the obligation.

This emendation will not, however, materially affect our argument.

Once again we shall have to defend our definition by an appeal to the role played by obligation in Hobbes's moral and political doctrines. We may find that men have obligations which cannot be derived from their own acts, restricting their initial rights. And indeed, one real difficulty does arise in defending our position.

We must deny that Hobbes appeals, or needs to appeal, to any obligations based on:

(i) an initial lack of natural right;
(ii) the laws of nature alone;
(iii) the will of God alone.

The third of these will occasion difficulties. Hobbes supposes that we have an obligation to obey God, and it is not easy to explain its precise character.

In this connection we may note that Hobbes occasionally uses 'obligation' or related terms in senses different from the moral sense we have characterized here. These uses occur primarily in *De Cive*. At one point Hobbes speaks of 'all we are obliged to by rational nature' (*E.W.* ii, p. 47), thus suggesting a concept of rational obligation which Oakeshott[1] tries to build into the core of Hobbes's theory. In a more important passage, Hobbes speaks of 'two species of *natural obligation*', distinguished from 'that *obligation* which rises from contract' (*E.W.* ii, p. 209). The first is clearly a non-moral or non-practical sense, for it is that in which 'heaven and earth, and all creatures, do obey the common laws of their creation' (ibid.). The latter, however, is the sense in which Hobbes argues we are obliged to obey God, and it will thus figure in our discussion of this problem, in chapter five.

D. *Justice*

Justice is perhaps the easiest of Hobbes's basic moral concepts to define formally. We need do no more than look at

[1] M. Oakeshott, *Introduction* to *Leviathan*, Oxford, 1955, p. lix.

Hobbes's statement, 'the definition of INJUSTICE, is no other than *the not performance of covenant*. And whatsoever is not unjust, is *just*' (*E.W.* iii, p. 131).

This gives us the equivalence:

'X is a just act' = 'X does not involve the not performance of covenant',

or more idiomatically:

'X is a just act' = 'X does not involve the breaking of covenant'.

The double negative is necessary; we cannot define justice as the keeping of covenants. For Hobbes supposes that acts are just, *unless* they involve the breaking of covenants, as his account makes clear.

Hobbes distinguishes just actions from just men. A just man may perform an unjust act, and conversely, an unjust man may perform a just act. To call a man just is to say that he is disposed to perform just acts; hence we may introduce the further equivalence:

'A is a just man' = 'A is disposed to perform acts which do not involve the breaking of covenant'.

These two equivalences provide a sufficient contextual formal definition of justice.

Hobbes does at one point suggest a slightly broader definition of justice. For he says:

And when a man hath . . . abandoned, or granted away his right; then he is said to be OBLIGED, or BOUND, not to hinder those, to whom such right is granted, or abandoned, from the benefit of it: . . . and that such hindrance is INJUSTICE, and INJURY, as being *sine jure*; the right being before renounced, or transferred. (*E.W.* iii, p. 119)

This passage suggests the equivalence:

'X is a just act' = 'X is an act which the performer has not laid down the right to perform'.

The previous equivalence would become merely a special case of this, covenant being one way of laying down the right to perform certain acts. But this wider definition is not in fact used by Hobbes, who generally speaks as if covenant were *the* way of laying down rights. We propose, therefore, not to advance this broader equivalence as Hobbes's formal definition of justice, but nothing of importance in Hobbes's argument turns on this point.

There is an interesting parallel between injustice and absurdity, developed most fully by Hobbes in *De Corpore Politico*:

> There is a great similitude between that we call *injury*, or *injustice* in the actions and conversations of men in the world, and that which is called *absurd* in the arguments and disputations of the Schools. For as he, which is driven to contradict an assertion by him before maintained, is said to be reduced to an absurdity; so he that through passion doth, or omitteth that which before by covenant he promised to do, or not to omit, is said to commit injustice; and there is in every breach of covenant a contradiction properly so called. For he that covenanteth, willeth to do, or omit, in the time to come. And he that doth any action, willeth it in that present, which is part of the future time contained in the covenant. And therefore he that violateth a covenant, willeth the doing and the not doing of the same thing, at the same time, which is a plain contradiction. And so *injury* is an *absurdity* of conversation, as absurdity is a kind of injustice in disputation. (*E.W.* iv, p. 96)

This passage raises certain difficulties for Hobbes's account of the will which we shall have to consider later in this chapter. Here we need note only that Hobbes does not argue that an act is unjust *because* it is contradictory. An act is unjust because it involves violation of covenant. But since to covenant is to will the action covenanted, and to violate covenant is to will an action incompatible with the action covenanted, injustice does involve a contradiction.[1]

Finally, it is worth noting the connection between injustice and the third law of nature, which in *Leviathan* is formulated simply: *'that men perform their covenants made'* (*E.W.* iii, p. 130). This law Hobbes terms 'the fountain and original of

---

[1] A. E. Taylor, op. cit., pp. 38–9, does come near to claiming that for Hobbes an act is unjust or wrong because willing it involves willing a contradiction. This is part of his attempt to develop parallels between Hobbes and Kant. I do not find the attempt convincing.

JUSTICE' (ibid.), and its connection with justice is indeed evident. But what is of interest to us here is that this third law of nature, unlike its fellows, is given a purely formal interpretation, independent of Hobbes's psychology. It also has a material interpretation, of course, which we shall be considering in the next part of this chapter.

The formal interpretation is suggested by the defence Hobbes gives of the law. For he says that without it 'covenants are in vain, and but empty words; and the right of all men to all things remaining, we are still in the condition of war' (ibid.). That is, unless men recognize themselves to be obliged by their covenants to perform what they have covenanted, then no real transfer of right has taken place, and hence no valid covenant has been made. A covenant is valid only if it transfers rights, but it transfers rights only if the parties cannot resume the rights at their pleasure. The parties cannot resume their rights only if they are obliged not to, but they are obliged not to only if they are so obliged by the covenant. Hence this argument reinforces our view, stated in the previous part of this section, that covenants oblige necessarily.

This concludes our exposition of the formal definitions of Hobbes's moral concepts. If our definitions prove acceptable, then we have succeeded in showing that Hobbes's moral concepts can be understood without reference to his psychological theory. But it is equally clear that no substantive conclusions can be drawn from the definitions we have given; Hobbes's moral and political doctrines are certainly not entailed by the concepts as defined here. To show how Hobbes reaches his conclusions we must turn to the material definitions.

But it is worth noting that certain formal consequences follow from our definitions. These have been stated previously, but it may be worth collecting together the most important ones.

1. The laws of nature do not impose limitations on the right of nature.

2. The right of nature is logically prior to all obligation.

3. All obligation is self-imposed, as a restriction on the right of nature.

4. The laws of nature do not create obligations.

All of these conclusions must be defended by an appeal to the role Hobbes's moral concepts play in his moral and political system.

2. THE MORAL CONCEPTS: MATERIAL DEFINITION

A. *The Right of Nature*

No difficulty arises in formulating the material definition of the right of nature. We need only combine the formal definition with Hobbes's claim that 'all the voluntary actions of men tend to the benefit of themselves; and those actions are most reasonable, that conduce most to their ends' (*E.W.* iii, p. 133) to obtain the preliminary material equivalence:

'A has the right to do X' M = 'A doing X is conducive to A's ends'.

Now since the principal end of a man is the conservation or preservation of his own nature, we may establish the further equivalence:

'A has the right to do X' M = 'A doing X is conducive to A's preservation'.

This is not quite sufficient. For it is A's belief that an action is conducive to his preservation, and not its actually being so conducive, that gives him the right to perform it, as Hobbes makes clear in *De Cive*, saying 'we have already allowed him to be judge . . . whether it doth or not' contribute towards his preservation 'insomuch as we are to hold all for necessary whatsoever he shall esteem so' (*E.W.* ii, p. 10). So we should phrase the equivalence:

'A has the right to do X' M = 'A doing X is (believed by A to be) conducive to A's preservation'.

We need now only add a qualification for natural right:

'A has the natural right to do X' M = 'A doing X is initially (believed by A to be) conducive to A's preservation'.

And this is but to reformulate Hobbes's original definition of the right of nature, quoted previously (*E.W.* iii, p. 116).

Now in the state of nature a man finds that 'there is nothing he can make use of, that may not be a help unto him, in preserving his life against his enemies', so that 'it followeth, that in such a condition, every man has a right to every thing; even to one another's body' (*E.W.* iii, p. 117). This is the derivation of Hobbes's famous right to all things. A man may do, or may possess, with right, anything whatsoever, *provided* it is conducive to his own preservation.

Hobbes's commentators have been vexed by the question whether this proviso limits the right of nature.[1] Clearly a man has the natural right to do whatever he considers conducive to his preservation. But does he have the right to do, first, what he considers contrary to his preservation, and second, what he considers indifferent to his preservation?

The first part of the question is easily answered. One cannot have reason to do what one believes will destroy oneself, since one has reason to do only what is conducive to one's ends, and preservation is the principal end. Hence one cannot have a right to do what one considers contrary to one's preservation.

But this does not mean that the right of nature is initially limited. For the question of right arises only within the framework of what is in principle possible. One neither has nor lacks the right to do what one cannot be motivated to do. And a normal man, in possession of his faculties, is necessarily motivated to preserve himself, and so cannot be motivated to destroy himself. As we have seen, the man who seeks to kill himself is, for Hobbes, 'not *compos mentis*'.

The second part of the question is more difficult. Hobbes seems to rule out the possibility of actions indifferent to

[1] Or not so vexed. See, e.g., Howard Warrender, op. cit., pp. 59–61, for the view that right is limited in the state of nature.

preservation in *De Cive* when he says that 'because whatsoever a man would, it therefore seems good to him because he wills it, and either it really doth, or at least seems to him to contribute towards his preservation, . . . it follows, that in the state of nature, to have all, and do all, is lawful for all' (*E.W.* ii, pp. 10–11). All actions performed in the state of nature, where the right of nature applies in full, are judged conducive to preservation, and so there can be no limitation on the right.

But Hobbes seems immediately to contradict this in a footnote: 'But if any man pretend somewhat to tend necessarily to his preservation, which yet he himself doth not confidently believe so, he may offend against the laws of nature, as in the third chapter of this book is more at large declared' (*E.W.* ii, p. 10 n.). Now the laws of nature, as we have interpreted them, indicate what is required by reason. To offend against them is to do what is contrary to reason. And one cannot do what is contrary to reason in accordance with the right of nature.

Hence Hobbes seems to be saying that the right of nature is not initially unlimited. A man might offend against the laws of nature in the state of nature, and such offences are not permitted by the right of nature. Note that this position is quite compatible with our claim that the laws of nature do not impose limitations upon the right of nature. For the question here is not whether the laws of nature are laws restricting our rights. Rather it is whether the laws of nature show the extent to which the right of nature is originally limited, by advising us that certain actions are wrong, contrary to reason.

I am, however, inclined to think that in the bare state of nature, the right of nature is strictly unlimited—that whatever we do in that state is to be taken as considered conducive to preservation, and so done with right. To reconcile the apparent conflict in Hobbes's statements we may note first the reference in the footnote to the third chapter. This may suggest to us that the conditions under which a man may offend against the laws of nature by doing what he only pretends to be conducive to preservation arise later, and are not to be supposed present in the original state of nature.

If this suggestion is accepted, then we may interpret the footnote, which is an addition to the original text introduced in the second Latin edition, as intended to anticipate an objection which the reader might raise in the light of subsequent arguments. Hobbes warns the reader not to suppose that because in the bare state of nature all actions are done with right, it follows that under all conditions all actions can be supposed to be intended to conduce to preservation, and so could be done with right.

To support this interpretation we must trace the course of the argument in *De Cive*. As we have already noted, this course differs from that in *Leviathan* in that the basic moral concepts are introduced prior to the analysis of conflict, terminating in the war of all against all. The argument proceeds as follows:

(1) If now to this natural proclivity of men, to hurt each other, . . . you add, the right of all to all, wherewith one by right invades, the other by right resists, . . . it cannot be denied but that the natural state of men, before they entered into society, was . . . a war of all men against all men. (*E.W.* ii, p. 11)

(2) But it is easily judged how disagreeable a thing to the preservation either of mankind, or of each single man, a perpetual war is. (*E.W.* ii, p.12)

(3) Whosoever therefore holds, that it had been best to have continued in that state in which all things were lawful for all men, he contradicts himself. For every man by natural necessity desires that which is good for him: nor is there any that esteems a war of all against all, which necessarily adheres to such a state, to be good for him. (Ibid.)

(4) Wherefore to seek peace, where there is any hopes of obtaining it, . . . is the dictate of right reason, that is, the law of nature; . . . (*E.W.* ii, p. 13)

(5) For if every one should retain his right to all things, . . . war would follow. He therefore acts against the reason of peace, . . . whosoever he be, that doth not part with his right to all things. (*E.W.* ii, p. 17)

In the beginning every man has an unlimited right to do what he will, conceiving it to be for his preservation. But the exercise of this unlimited right is one of the causes of the war of all against all, which is inimical to preservation. Thus the unlimited right of nature proves contradictory in its use; the man who exercises his right in order to preserve himself contributes thereby to the war of all against all, which tends to his own destruction. And so it is necessary to give up some part of the unlimited natural right.

But a particular action, forbidden by the limitations accepted on the right of nature, may still seem desirable to a man. And so he may pretend that it conduces to his preservation, though he cannot confidently believe so. If he acts on this pretence, he acts against what is necessary to peace, and so against reason. He acts without right—for he goes beyond the limitations accepted on the right of nature.

By this argument, then, we can reconcile the apparently contradictory passages we quoted; initially, the right of nature is unlimited, but after one has accepted limitations on it, disregard of these limitations is without right, and indeed contrary to the laws of nature. And this argument also introduces in bald outline the exposition of our next section, for it suggests the material definition and content of the laws of nature.

B. *Law of Nature*

The material definition of a law of nature follows directly from Hobbes's own account (*E.W.* iii, pp. 116–17), although it could also be constructed from the formal definition, by substituting for 'the requirements of right reason' a phrase indicating the actual content of those requirements. We arrive at the following equivalence:

'law of nature' M = 'precept laying down what is required
   for preservation'.

The content of the laws of nature is Hobbes's positive moral doctrine. These precepts, derived from his account of human nature, lay down what is necessary if man is to achieve, as far as is possible, his principal end—his own preservation, and with it, his well-being.

The fundamental law of nature is '*that every man, ought to endeavour peace, as far as he has hope of obtaining it'*, or more simply, it is *'to seek peace, and follow it'* (*E.W.* iii, p. 117). This law is the most general conclusion man derives from his experience of the war of all against all. Clearly it depends on that experience, whether real or imagined. Although hypothetically a man might conclude that war was necessarily inimical to human life, only an analysis of the human condition with all social bonds removed shows that peace is the *primary* requisite for preservation.

This precept tells us what to do, to better our chances of survival. It does not, then, limit our right of nature, but rather prescribes how to employ that right. Of course, action against this precept would be wrong. But such action can consist only in seeking war, rather than peace, and Hobbes insists both that no one 'esteems a war of all against all . . . to be good for him' and that 'every man by natural necessity desires that which is good for him' (*E.W.* ii, p. 12). It follows that no man, or no sane man, wittingly violates this precept, in so far as he understands it.

As we saw in the previous section, Hobbes holds that as long as men possess and act on the unlimited right of nature, war must result. Hence:

From this fundamental law of nature, by which men are commanded to endeavour peace, is derived this second law; *that a man be willing, when others are so too, as far-forth, as for peace, and defence of himself he shall think it necessary, to lay down this right to all things; and be contented with so much liberty against other men, as he would allow other men against himself.* (*E.W.* iii, pp. 117–18)

The second law of nature, then, provides the rationale for laying down some part of the initially unlimited right of nature, which is to say, for taking on certain obligations. Like the first law, it is a rational precept, telling us how best to secure our preservation. In itself it does not limit the exercise of the right of nature, BUT it tells us to impose such a limit. Thus in acting on it we exercise our previously unlimited right of nature, to perform an action which has the effect of limiting our right of nature.

It is essential, if we are to understand the role of the laws of nature and their relation to obligation and to the right of nature, to be clear about the exact import of this second law. It does *not* state that as a condition of peace every man's right of nature is limited in certain respects. If it did, then our claim that the laws of nature do not impose limitations on the right of nature would be flatly false.

It *does* state that as a condition of peace every man must limit his own right of nature in certain respects, provided others do so. In this way, it does *indirectly* provide for limitations on the right of nature, by imposing on us the rational requirement that we *directly* limit our right.

We may express this crucial distinction in another way. We do not limit our right of nature by rationally establishing the second law of nature. But once we have established it, we do limit our right of nature by acting in accordance with it. The limitation derives from our action and not from the law, but our action is required by the law.

Promising, as we ordinarily understand it, provides an analogy. We do not impose obligations on ourselves by deciding that we ought to make certain promises.[1] We do impose obligations by actually making the promises. The obligation derives from our act of promising, and not from the judgement that we ought to promise, but our act is required by the judgement.

It is understandably easy to misconceive the role of Hobbes's second law of nature. Confusing what the law does with what it requires us to do, one then supposes that the law does impose restrictions on the right of nature. And then, supposing that 'obligation' means merely 'lack of right', one supposes that the law directly obliges us. And then one must deny that all obligation is self-imposed. In this way three of the four formal consequences we drew from our formal definitions of Hobbes's moral concepts are rejected. If it is further supposed

---

[1] I assume here that ordinary 'ought'-judgements are not themselves judgements of obligation. I *ought* to get on with writing this morning, but I do not have an *obligation* to do so.

that the second law of nature imposes an initial restriction on the right of nature, obliges us *ab initio*, then the remaining formal consequence (the second one) is also rejected.

Now it would be disingenuous not to admit that Hobbes himself is not entirely free from confusion with respect to the role of the second law of nature. In particular, he does speak of the laws of nature as obliging, and we shall have to consider in the next section of the chapter whether what he says can be reconciled with our definitions and their consequences. At this point let me claim only that our account of the role of the second law of nature is consistent with what Hobbes explicitly and specifically says in introducing the law, and that our interpretation enables us to ascribe to Hobbes a consistent doctrine, that various parts of this doctrine are explicitly stated in important passages in Hobbes's works, that the doctrine provides the conclusions Hobbes draws as far as these conclusions are consistent, and that it is Hobbesian in spirit. And by that last, I mean that it is a rational doctrine for self-interested, secular, rational men.

The third law of nature, *'that men perform their covenants made'* (*E.W.* iii, p. 130), has already been introduced into our discussion. We examined its formal interpretation in the first part of this chapter; here we are concerned with its material interpretation—that is, with its role as a condition of peace.

Hobbes defends this law on this interpretation in his discussion of it in *De Cive*:

> For it hath been showed in the foregoing chapter, that the law of nature commands every man, as a thing necessary, to obtain peace, to convey certain rights from each to other; and that this . . . is called a contract. But this is so far forth only conducible to peace, as we shall perform ourselves what we contract with others shall be done or omitted; and in vain would contracts be made, unless we stood to them. Because therefore to stand to our covenants, or to keep faith, is a thing necessary for the obtaining of peace; it will prove . . . to be a precept of the natural law. (*E.W.* ii, pp. 29–30)

The second law tells us that granting away rights—making covenants—is necessary to peace; the third law adds that

keeping covenants is necessary to peace. One might think
this a pointless addition, since if keeping covenants were not
necessary to peace, then men would not keep covenants,
and so making covenants would be pointless, and could not
then be necessary to peace. We shall find, however, that this
argument is too simple; there is a point to distinguishing
the second and third laws, and a point which will give Hobbes
trouble. For there is always the man who reasons that,
although it may be necessary for everyone to make covenants
to secure peace, it is enough if only most people keep their
covenants, so that he may break his.

The third law is naturally appealed to by those who interpret Hobbes as maintaining that the laws of nature do
oblige man directly, and that covenants do not oblige of themselves. For, it is urged, why would Hobbes introduce this
third law, if not as an answer to the question 'why am I
obliged to keep my covenants?'

But our discussion shows that Hobbes has good reason for
introducing this law. Formally interpreted, it reinforces the
claim that covenants oblige of themselves. Materially interpreted, it affirms that keeping covenants, as well as making
them, is a condition of peace and so of preservation. Both
these considerations are found in Hobbes's discussion of the
law. What is not found is the claim that the law is the ground
of our obligation to keep covenants.

Hobbes enumerates several further laws of nature—six-
teen, to be exact, in *Leviathan*—most of which need not
concern us. They make no substantial contribution to his
moral doctrine. We should, however, note the fourth law,

*that a man which receiveth benefit from another of mere grace, endeavour
that he which giveth it, have no reasonable cause to repent him of his good
will. . . . The breach of this law, is called* ingratitude; *and hath the same
relation to grace, that injustice hath to obligation by covenant.* (*E.W.* iii,
p. 138)

'Grace' is an equivalent for 'gift' or 'free-gift' (*E.W.* iii,
p. 121). Hobbes claims in this fourth law that showing gratitude for gifts received is a condition of peace. But he never

suggests that men are *obliged* to display gratitude[1]—as he should if the laws of nature were grounds of obligation. My right of nature is not restricted should you confer some kindness upon me.

The ninth and tenth laws impose, as conditions of peace, that men acknowledge each other as equal, and accept equal rights in establishing conditions of peace (*E.W.* iii, pp. 140–1). The man who demands more for himself than he allows to others is arrogant. The arrogant man does not violate his obligations; instead, he refuses to accept obligations which he insists others must accept as a condition of peace. In this way he prevents the attainment of peace, and so, although he is not unjust, as an enemy to peace he may be killed by others—without injustice.

After listing those laws of nature which are conditions of peace, Hobbes mentions 'other things tending to the destruction of particular men; as drunkenness, and all other parts of intemperance; which may therefore also be reckoned amongst those things which the law of nature hath forbidden; but are not necessary to be mentioned, nor are pertinent enough to this place' (*E.W.* iii, p. 144). Hobbes often speaks as if he equated the laws of nature with the conditions of peace. But his general definition does not contain this restriction, which would rule out precepts such as that against drunkenness. However, these further laws of nature play no significant role in Hobbes's moral theory, and none at all in his political theory. Perhaps we might label them rules of personal prudence.

C. *Obligation*

To lay down a right is to undertake an obligation. In the previous section we have outlined the rationale for laying down rights, and hence for assuming obligations. Although Hobbes does not in fact provide a material definition for the concept of obligation, the obvious way to construct the

---

[1] Howard Warrender, op. cit., p. 52, speaks of an obligation to show gratitude. But the text gives no support for this.

definition is to add the material condition for laying down rights to the formal equivalence. Thus we obtain:

'A has an obligation not to do X' M = 'A has laid down the natural right to do X, as a condition of preservation',

or alternatively:

'A has an obligation not to do X' M = 'A has laid down the natural right to do X, in accordance with the second law of nature'.

We have already noted that covenant is the usual means by which men oblige themselves. But simple renunciation or transfer of right is at least theoretically possible. The difference between renunciation and transfer is this: if I renounce my right to some object, I oblige myself not to impede any man in the exercise of his right to that object; if I transfer my right to some object, I oblige myself only not to impede the person to whom I transfer the right in the exercise of his right to that object (*E.W.* iii, pp. 118–19).

Not all rights, or all parts of the right of nature, can be laid down, and hence there are limits on the possible extent of obligation. One cannot give up one's right to do what is immediately and directly necessary for preservation—to resist threats and assaults on one's life, power, and corporal liberty (*E.W.* iii, p. 120). There is, then, an inalienable core to the right of nature. But note again that, although I cannot have an obligation not to defend myself, this of itself imposes no obligation on others not to attack me.

In addition to this primary restriction on the extent of our obligations, there are certain secondary restrictions on the circumstances under which we may be obliged, and the objects to which we may be obliged, which arise from limitations on the validity of covenants.[1] Only two of these restrictions are of importance to our argument. 'To make covenant with God, is impossible, but by mediation of such as God speaketh

---

[1] Howard Warrender, op. cit., chapter iii, has a full and excellent discussion of the conditions under which covenants are invalid.

to, either by revelation supernatural, or by his lieutenants that govern under him, and in his name: for otherwise we know not whether our covenants be accepted, or not' (*E.W.* iii, pp. 125–6).

This restriction will affect the nature of our obligation to obey God.

The second restriction bears directly on Hobbes's political theory. For Hobbes supposes that we escape from the state of nature, and enter political society, by covenant, but he can be interpreted as holding that covenants made in the state of nature are invalid. And these positions are inconsistent, for if covenants made in the state of nature are invalid, no covenant can be made to get us out of the state of nature.

But Hobbes's real position does not involve this inconsistency. He says:

If a covenant be made, wherein neither of the parties perform presently, but trust one another; in the condition of mere nature, which is a condition of war of every man against every man, upon any reasonable suspicion, it is void: . . .

The cause of fear, which maketh such a covenant invalid, must be always something arising after the covenant made; . . . For that which could not hinder a man from promising, ought not to be admitted as a hindrance of performing. (*E.W.* iii, pp. 124–5)

Covenants of mutual trust are, then, precariously valid in the state of nature. Either party, acting in good faith, may void such a covenant by indicating some fact, posterior to the making of the covenant, as ground for suspecting the other party of non-fulfilment. But we need not suppose this always and inevitably to happen. And so men are not prevented, in the state of nature, from making that covenant which terminates the state of nature, and begins political society.

We have now sufficiently demarcated the extent of our obligations. We have an obligation to do whatever we have renounced the right to do, in the interests of our own preservation. We may have obligations in the state of nature, although they hold only precariously. And so the obligations we undertake in the state of nature are not sufficient to end the war of all against all, and bring about a state of security for all men. Thus acting on the second law of nature, although necessary for our preservation, is not sufficient to secure our preservation.

We should also consider briefly the character of our obligations. Hobbes supposes that all obligations are undertaken on prudential grounds—as means to self-preservation. But this does not make obligation a prudential concept. To say that obligations are undertaken for prudential reasons is not to say that we are under obligation only so long as we find it to our interest.

Again we may compare promising. Very often we make promises on strictly prudential grounds. I promise to do you some favour in return for your performing some action on my behalf. I make this promise strictly and solely because it is in my interest to do so, because I stand to benefit from what you do on my behalf, and only my promise will induce you to do it. But I am not therefore under obligation only so long as I find it advantageous. Whether or not I do benefit from your action I have an obligation to do you the promised favour.

This point has not been sufficiently appreciated by some of Hobbes's commentators. Thus Warrender argues that if the laws of nature are considered to be only 'prudential maxims for those who desire their own preservation', then 'political obligation would turn out like natural law to be no more than another prudential maxim'.[1] But the conclusion does not follow. From the fact that the laws of nature are prudential maxims, telling us to put ourselves under obligation, it in no way follows that obligation itself is only a prudential maxim.

But there is, it must be admitted, a difficulty here for Hobbes. We may suppose it possible to oblige ourselves on prudential grounds without taking obligation to be itself prudential, because we suppose that we may have non-prudential reasons for fulfilling our obligations. If not, then if no advantage would accrue to us from fulfilling the obligation, we could not suppose we had any reason to do so, and this would

[1] Howard Warrender, op. cit., pp. 99–100.

divest the claim that we had an obligation of its usual significance.

Now Hobbes does not suppose that we have non-prudential
reasons for acting. Practical reason is linked to self-preservation. Of course, as the third law of nature makes clear, he
does maintain that we have prudential reasons for fulfilling our
obligations. But whether he then has a moral theory of obligation is a question we shall consider in the final part of this
chapter.

D. *Justice*

As with obligation, Hobbes does not provide a material
definition of the concept of justice. But the construction of one
is easily carried out, using the same method as before.

'X is a just act' M = 'X does not involve the breaking of a
covenant undertaken in accordance with the second law
of nature'.

We need not state the similar equivalence for 'A is a just
man'.

Hobbes raises, in connection with justice, a problem which
has an important bearing on the question raised about obligation in the last section. A rather lengthy quotation will serve
to introduce it:

The fool hath said in his heart, there is no such thing as justice; . . .
seriously alleging, that every man's conservation, and contentment,
being committed to his own care, there could be no reason, why every
man might not do what he thought conduced thereunto: and therefore
also to make, or not make; keep, or not keep covenants, was not against
reason, when it conduced to one's benefit. He does not therein deny,
that there be covenants; and that they are sometimes broken, sometimes
kept; and that such breach of them may be called injustice, and the
observance of them justice: but he questioneth, whether injustice,
taking away the fear of God, for the same fool hath said in his heart
there is no God, may not sometimes stand with that reason, which
dictateth to every man his own good; . . . From such reasoning as this,
successful wickedness hath obtained the name of virtue: . . . you may
call it injustice, or by what other name you will; yet it can never be

against reason, seeing all the voluntary actions of men tend to the benefit
of themselves; and those actions are most reasonable, that conduce most
to their ends. (*E.W.* iii, pp. 132–3)

The problem is clear. If I make a covenant, then it would be
unjust, and contrary to obligation, to break it. But if it is to my
advantage, then it cannot be contrary to reason. And so the
third law of nature is not a precept of reason. Neither obligation nor justice need accord with reason.

But there is worse to come. What I do with reason, I do
with right. If reason dictates the breaking of covenant, showing it to be to my benefit, then I must have the right to do
what is contrary to the covenant. But in making the covenant
I renounced that right. Therefore I have the right to do what I
have renounced the right to do—I have the right to do what
I have an obligation not to do. And with this, the Hobbsian
moral system collapses in inconsistency.

'This specious reasoning is nevertheless false' (*E.W.* iii,
p. 133). Hobbes's arguments must await the next part of our
study where they can be considered as part of a total discussion of the bearing of motivation on obligation in Hobbes's
theory.

3. THE MORAL CONCEPTS: SOME PROBLEMS

A. *Liberty*

We postponed discussion of Hobbes's conception of liberty,
despite its appearance in his definition of the right of nature,
because any earlier attempt to untangle the confusions Hobbes
falls into with liberty would have impeded and confused our
own analysis. Having now defined Hobbes's principal moral
concepts, and determined the consequences, both formal and
material, of our definitions, we may return to an assault on
liberty.

Hobbes gives two definitions of the concept of liberty in
Leviathan. The first follows immediately his account of the
right of nature; the second occurs when he comes to consider
the liberty of subjects in civil society. They are:

By LIBERTY, is understood, according to the proper signification of the word, the absence of external impediments: which impediments, may oft take away part of a man's power to do what he would; but cannot hinder him from using the power left him, according as his judgment, and reason shall dictate to him. (*E.W.* iii, p. 116)

LIBERTY, or FREEDOM, signifieth, properly, the absence of opposition; by opposition, I mean external impediments of motion; and may be applied no less to irrational, and inanimate creatures, than to rational. . . . But when the impediment of motion, is in the constitution of the thing itself, we use not to say; it wants the liberty; but the power to move; . . . (*E.W.* iii, p. 196)

These passages are not quite consistent. For the first suggests that liberty and power are coextensive, that what a man is free to do is what he has power to do. But the second makes a distinction between them, which is supported elsewhere:

It is one thing to say a man hath liberty to do what he will, and another thing to say he hath power to do what he will. A man that is bound, would say readily he hath not the liberty to walk; but he will not say he wants the power. But the sick man will say he wants the power to walk, but not the liberty. (*E.W.* v, p. 265)

But if power is a man's 'present means; to obtain some future apparent good' (*E.W.* iii, p. 74), then surely external impediments do take away a man's power, and the man who is bound, just as the man who is sick, must want the power to walk. To make Hobbes's account of liberty and power consistent, we must suppose that power is considered in three different ways.

First, there is potential power, or capacity. A man or a thing has the capacity to move in some way, if and only if it could move in that way under physically possible external conditions. Thus man has the capacity to walk, but not to fly. In this sense the well man, the sick man, and the bound man, all have the power to walk.

Second, there is actual power. A man or a thing has the actual power to move in some way, if and only if it can now move in that way under physically possible external conditions. In this sense the well man and the bound man, but not the sick man, have the power to walk.

Third, there is present power. A man or a thing has the present power to move in some way, if and only if it can now move in that way under actual external conditions. In this sense the well man, but neither the bound man nor the sick man, has the power to walk.

Liberty, now, must be defined as potential power unimpeded by external conditions. A man or a thing has the liberty to move in some way, if and only if it could now move in that way under actual external conditions. Thus the well man and the sick man, but not the bound man, have the liberty to walk.

With a little forcing, this definition will serve to explain Hobbes's account of deliberation. For he says: 'And it is called *deliberation*; because it is a putting an end to the *liberty* we had of doing, or omitting, according to our own appetite, or aversion' (*E.W.* iii, p. 48).

Deliberation terminates in action. But when we act, then we are no longer at liberty to act or not to act. Although our capacities are unchanged, yet the 'external circumstances' of our acting prevent us from choosing between acting or not.

Note that this explanation requires that action follows immediately on deliberation. If it did not, then our liberty would not be at an end, for the actual external circumstances would still permit us to exercise our capacity to choose— assuming that the circumstances permitted this initially. And Hobbes insists, in two very important passages, that action must follow immediately on deliberation and its terminus, will.

I conceive that in all *deliberations*, that is to say, in all alternate *succession* of contrary *appetites*, the last is that which we call the WILL, and is immediately next before the doing of the action, or next before the doing of it become impossible. All other *appetites* to do, and to quit, that come upon a man during his deliberations, are called *intentions* and *inclinations*, but not *wills*, there being but one *will*, which also in this case may be called the *last will* . . . (*E.W.* iv, p. 273)

My opinion is no more than this, that a man cannot so determine to-day, the will which he shall have to the doing of any action to-morrow, as that it may not be changed by some external accident or other, as there shall appear more or less advantage to make him persevere in the will to the same action, or to will it no more. (*E.W.* v, p. 209)

Now let us draw the consequences of our discussion of Hobbes's concept of liberty. First, it must be apparent that we have utterly misconceived Hobbes's account of the right of nature. We must now say, following Hobbes's definition (*E.W.* iii, p. 116), that the following equivalence holds:

'A has the right to do X' M = 'A has the liberty to do X, as a means to A's preservation';

which becomes:

'A has the right to do X' M = 'A has the capacity, unimpeded by actual external conditions, to do X, as a means to A's preservation'.

The right of nature, it turns out, has nothing whatsoever to do with what a man *may* do. It is a purely descriptive concept, stating what he *can* do, in order to preserve himself.

Second, consider the effect of this doctrine of liberty on our account of covenant. Hobbes says: 'Promises therefore, upon consideration of reciprocal benefit, are covenants and signs of the will, or last act of deliberation, whereby the liberty of performing, or not performing, is taken away, and consequently are obligatory. For where liberty ceaseth, there beginneth obligation' (*E.W.* iv, pp. 90–1).

Since liberty is terminated by covenant, then the covenant must remove either the covenanters' capacity not to perform or, more likely, must erect external impediments in the way of their not performing. It follows, then, that after covenanting, a man is no longer able not to perform. Away, then, with the third law of nature! Away with the distinction Hobbes confusedly makes between '*natural obligation* . . . when liberty is taken away by corporal impediments' and 'that *obligation* which rises from contract' (*E.W.* ii, p. 209), for the latter reduces to a species of the former! Away, indeed, with all *moral* terms whatsoever!

It will be evident that something has gone astray. And the obvious candidate is Hobbes's concept of liberty. What we require, in addition to the sense of liberty we have defined, is a quite different sense, in which liberty is related to what we *may* do, not what we *can* do. And we shall have to distinguish also those deliberations which terminate in decisions to act, which are immediately effected, from those which terminate in decisions to oblige oneself to act in the future. The former will terminate what we may call *physical* liberty; the latter will terminate *moral* liberty.

It might be supposed that our second sense of liberty could embrace both 'may' and 'can'. But this will not make sense of Hobbes's account of natural right. Natural right, it will be remembered, is originally unlimited. And Hobbes contends: 'For he that renounceth, or passeth away his right, giveth not to any other man a right which he had not before; because there is nothing to which every man had not right by nature: but only standeth out of his way, that he may enjoy his own original right, without hindrance from him; . . .' (*E.W.* iii, p. 118).

If I enable you to do or obtain something, which previously I had prevented you from doing or obtaining, then I do remove an external impediment to your power. You are now free from my hindrance. But, Hobbes insists, this affects not your right but your enjoyment of it. Thus your right is not limited by external impediments, and so extends beyond your physical liberty.

Thus moral liberty, which appears in Hobbes's account of the right of nature, which is contrasted with obligation, and which is terminated by covenant, must be quite independent of physical liberty. Indeed, moral liberty does not seem to be a separate concept from right, in Hobbes's argument, and so cannot serve to explain the right of nature. Our excursion into Hobbes's conception of liberty has added nothing to our previous statement of his position.

B. *Laws of Nature and Obligation*

No doubt the most controversial of the consequences we drew from our formal definitions is that the laws of nature do not create obligations. As we have explained, men create obligations for themselves by acting on the second law of nature. But the laws of nature are not themselves obligatory.

This consequence follows from our definition of the laws as rational precepts. Now we have suggested that Oakeshott finds in Hobbes a sense of obligation which he denominates *rational*, and contends that rational precepts ' "oblige" on account of their rationality'.[1] Oakeshott explains rational obligation by saying that 'a man may be prevented from willing a certain action because he perceives that its probable consequences are damaging to himself. Here the impediment is internal, a combination of rational perception and fear, which is aversion from something believed to be hurtful.'

Although there is only dubious justification in Hobbes's writings for erecting this concept of rational obligation, we have no major quarrel with Oakeshott, if he wishes to suppose that rational precepts are rationally obliging. For as he admits, 'the natural Right to act in any way he chooses has suffered no impediment; fear and reason may limit a man's power, but not his Right'. Thus rational obligation is not obligation in the sense we have defined; Oakeshott's position is consistent, then, with our claim that the laws of nature do not create moral obligations, and indeed with the claim that the laws of nature do not impose limitations on the right of nature.

The primary justification we would offer for our claim that the laws of nature do not create moral obligations—using the word 'moral' in the broad sense of 'practical'—is our success in explaining how Hobbes is able to introduce obligation into his system without appealing to the laws of nature as initially obligatory. The role played by the laws of nature does not require them to be laws, and does not require them to be obligatory. As long as they serve as rational precepts, telling men to put themselves under obligation, Hobbes is able to construct his system.

Compare the economy of this account with the difficulties inherent in supposing that the laws of nature are obligatory. If obligatory, they must be genuine laws. And as laws, they

---

[1] Michael Oakeshott, op. cit., all quotations from p. lix.

must be commands, 'nor a command of any man to any man; but only of him, whose command is addressed to one formerly obliged to obey him' (*E.W.* iii, p. 251). The laws of nature must then be considered the commands of God, and their obligatory force must be derived from a prior—and presumably ultimate—obligation on men, to obey God.

But nothing in Hobbes's argument, addressed to men conscious of the perils of the state of nature and anxious to find security, depends on this appeal to laws of nature as commands of God whom one is ultimately obliged to obey. A man who seeks to preserve himself, and who agrees that the laws of nature are necessary to preservation, needs no such appeal to convince him. And in Hobbes's view every sane man seeks to preserve himself, and every rational man will agree that the laws of nature are necessary to his preservation.

Hobbes says that 'law in general, is not counsel, but command' (ibid.). This is an important distinction, worth pursuing in connection with the laws of nature. Counsel and command share the grammatical form of an imperative: 'For the words *do this*, are the words not only of him that commandeth; but also of him that giveth counsel; and of him that exhorteth; . . .' (*E.W.* iii, p. 240). Thus the fact that we find the laws of nature expressed as imperatives does not show them to be commands, rather than counsels.

> COMMAND is, where a man saith, *do this*, or *do not this*, without expecting other reason than the will of him that says it. From this it followeth manifestly, that he that commandeth, pretendeth thereby his own benefit: for the reason of his command is his own will only, and the proper object of every man's will, is some good to himself.
>
> COUNSEL, is where a man saith, *do*, or *do not this*, and deduceth his reasons from the benefit that arriveth by it to him to whom he saith it. And from this it is evident, that he that giveth counsel, pretendeth only, whatsoever he intendeth, the good of him, to whom he giveth it. (*E.W.* iii, p. 241)

The laws of nature, as we have seen, are deduced from the benefit of those to whom they are delivered. In this respect, clearly, they resemble counsel more closely than command.

Now Taylor observes that 'Hobbes always describes the items of the natural law as *dictamina*, or dictates, never as *consilia*, or pieces of advice'.[1] We must not identify the laws of nature with mere counsel. But this is compatible with our view that the laws resemble counsels.

Good grounds for not treating the laws of nature as counsels are easily found. In the first place, counsel implies a counsellor; once again divine origin for the laws of nature would be required. Yet such origin plays no part in the role assigned to the laws in Hobbes's argument. There, the laws are derived from reason alone. But, in the second place, it is more natural to speak of the dictates of reason than the counsels of reason. And in the third place, and most important, the laws of nature are 'theorems'; they are necessary conclusions of reason in a way in which mere advice is not.

Perhaps we might term the laws of nature *prescriptions*. A prescription is rather more than mere counsel, though it does not have the force of a command. Such a term would help to clarify the precise role which, in our view, the laws of nature play in Hobbes's argument.

In claiming that the natural laws are not laws, we are not of course denying that in some contexts they become laws. In a passage which strongly supports the position we are urging, Hobbes says:

> For the laws of nature, . . . are not properly laws, but qualities that dispose men to peace and obedience. When a commonwealth is once settled, then are they actually laws, and not before; as being then the commands of the commonwealth; and therefore also civil laws: for it is the sovereign power that obliges men to obey them. (*E.W.* iii, p. 253)

In civil society the laws of nature become civil laws. But it is evident that it is not as civil laws that they enter into Hobbes's moral theory.

Furthermore, the laws of nature are of course regarded by Hobbes as divine laws, 'delivered by God in holy Scriptures' (*E.W.* ii, p. 50). And as divine laws they oblige, in virtue of our obligation to obey God. We do not deny that Hobbes

[1] A. E. Taylor, op. cit., p. 40.

supposes we have such an obligation. Our view is that it is not as divine laws, obliging as commands of God, that the laws of nature enter into Hobbes's moral theory.

Hopefully, our position should be clear. We must now examine the obvious objections which may be advanced against it. First, we must consider why, if we are right, Hobbes speaks of the *laws* of nature. For on our view this label is misleading; surely Hobbes would not have confused his readers in this way had he really supposed the laws of nature not to be laws. And second, there are passages in Hobbes's several treatises in which he specifically speaks of the laws of nature as obliging. How, on our view, are these passages to be explained?

Hobbes speaks of the laws of nature because he has not fully emancipated himself from the medieval conception of natural law. He is not fully aware of his own originality, and hence he fails to see that the language of medieval legal and political thought undergoes a sea-change within the framework of his own theory.

Very briefly, the change is this. In medieval thought, natural law is God's law for the governance of mankind, known through natural reason. The practical role of reason— or the role of practical reason—is to discern this law; thus the law of reason and the divine law are necessarily connected.

Hobbes retains the term 'natural law', and its equation, both to divine law and to the practical precepts discerned by reason. But for Hobbes the practical role of reason is to establish the means of preservation which are the conditions of peace, and not to discern the rationally unknowable will of God. Thus the dictates of reason and the laws of God are no longer necessarily connected; if God wills man's preservation, then the two will coincide, but this will require independent demonstration. And the dictates of reason are not in themselves law; only *if* they are commanded by God are they *also* laws.

What is important to Hobbes's moral and political theory is natural law *qua* dictate of reason, not *qua* command of God. Hence Hobbes's label, law of nature, misleads in containing 'law'. But only we, who are the beneficiaries of the revolution in moral and political concepts which Hobbes, among others, initiated, are in a position to draw clearly the conceptual lines with which Hobbes struggled.

We may add further that God, as author of the universe, must be author also of the conditions of our preservation. In this sense, then, we may think of the precepts stating those conditions as part of the natural law of God, though 'law' here will have lost its connection with 'command'. The laws of nature will be natural prescriptive principles in much the sense that the laws of physics are natural descriptive principles. Hobbes's use of the label, law of nature, no doubt reflects these considerations as well.

We turn now to those passages in which Hobbes actually speaks of the laws of nature as obliging. These passages may conveniently be divided into three main groups, the first of which concern our obligation to keep covenants.

> For the law of nature . . . commands us to *keep contracts*; and there-fore also to perform obedience, when we have covenanted obedience, . . . But all subjects . . . do *covenant* to obey his commands who hath the supreme power, that is to say, the civil laws, in the very constitution of government, even before it is possible to break them. For the law of nature did oblige in the state of nature; . . .
> Seeing therefore our obligation to observe these laws is more ancient than the promulgation of the laws themselves, as being contained in the very constitution of the city; by the virtue of the natural law which forbids breach of covenant, the law of nature commands us to keep all the civil laws. (*E.W.* ii, p. 190)
> For a civil law, that shall forbid rebellion, . . . is not, as a civil law, any obligation, but by virtue only of the law of nature, that forbiddeth the violation of faith; which natural obligation, if men know not, they cannot know the right of any law the sovereign maketh. (*E.W.* iii, pp. 323–4)
> Whereas signs and miracles had for end to procure faith, not to keep men from violating it, when they have once given it; for to that men are obliged by the law of nature. (*E.W.* iii, pp. 469–70)

We might summarize our interpretation of the laws of nature, as they apply to keeping covenants, or more generally, keeping faith, in three propositions:

1. As a condition of peace, we must make covenants and, more generally, lay aside certain rights.
2. As a condition of peace, we must keep covenants and, more generally, keep faith, or hold to our laying aside of rights.
3. Valid covenants and, more generally, layings aside of rights, oblige.

It is tempting to sum this up in the claim that the laws of nature oblige us to the keeping of faith, even if it is not strictly accurate. And we have no reason to expect Hobbes, whose primary interest is in drawing moral and political conclusions and not in making conceptual distinctions, to have noted or cared about the inaccuracy. What we must determine is whether Hobbes's conclusions are affected, if we replace his actual words by what we claim is a more accurate statement. And clearly his conclusions are not affected. Our obligation to keep civil laws is prior to the laws themselves, and found in the covenant which constitutes the city. Our obligation not to rebel also stems from that covenant. And we make that covenant in fulfilling the rational requirements of the law of nature.

Our interpretation, then, grants Hobbes all he needs or uses to draw his conclusion, and as we have said, grants it without also saddling him with the regress back to the obligation to obey God. Thus we do not find in these passages any real objection to our interpretation.

The second group of passages concerns the obligations of the sovereign.

But if there be none that can give the sovereignty, after the decease of him that was first elected; then has he power, nay he is obliged by the law of nature, to provide, by establishing his successor, to keep those that had trusted him with the government, from relapsing into the miserable condition of civil war. (*E.W.* iii, p. 179)

The OFFICE of the sovereign, . . . consisteth in the end, for which he was trusted with the sovereign power, namely the procuration of *the safety of the people*; to which he is obliged by the law of nature, and to render an account thereof to God, the author of that law, and to none but him. (*E.W.* iii, p. 322)

As the second passage makes clear, the sovereign's obligation to obey the law of nature arises from considering it as the law of God. In Hobbes's theory the law of nature, as it applies to individual men in the state of nature, is the dictate of reason, but, as it applies to sovereigns, is divine law. Although Hobbes insists that the interest and preservation of the sovereign require him to conform to the law of nature, he recognizes that this interest speaks less plainly to one in a position of secure power, than to one embroiled in the war of all against all. Hence he emphasizes the subordination of the sovereign to God as the only effective check upon the sovereign's conduct.

This does not affect our argument. For it is the application of the law of nature to men in the state of nature that determines the basic content of Hobbes's moral and political system. If they, acting in accordance with the precepts of that law, did not create Leviathan and set some man or group in a position of sovereign power, there would be no sovereign to whom the law of nature, in its secondary aspect as the law of God, could be applied.

The final, and most difficult group of passages, occurs at the conclusion of Hobbes's outline of the several laws of nature. He turns to consider the manner in which they are to be put into effect, and in *Leviathan* his remarks are given this marginal summary: 'The laws of nature oblige in conscience always, but in effect then only when there is security.' In the text itself, he says, 'The laws of nature oblige *in foro interno*; that is, to say, they bind to a desire they should take place: but *in foro externo*; that is, to the putting them in act, not always.' And he later says, 'The same laws, because they oblige only to a desire, and endeavour, I mean an unfeigned and constant endeavour, are easy to be observed' (*E.W.* iii, p. 145).

Before considering how these passages are to be interpreted, we should note that the same matter is discussed both in *De Corpore Politico* and in *De Cive*. In the former no mention is made of obligation, although both 'in foro interno' and 'in foro externo' appear (*E.W.* iv, pp. 107–8). In *De Cive*, on the other hand, obligation is emphasized, but in contexts which suggest that rational obligation is in question. The key passage is:

It is not therefore to be imagined, that by nature, that is, by reason, men are obliged to the exercise of all these laws in that state of men wherein they are not practised by others. We are obliged yet, in the interim, to a readiness of mind to observe them, whensoever their observation shall seem to conduce to the end for which they were ordained. We must therefore conclude, that the law of nature doth always and everywhere oblige in the internal court, or that of conscience; but not always in the external court, but then only when it may be done with safety. (*E.W.* ii, pp. 45–6)

Now if we may treat obligation *in foro interno* and *in foro externo* as species of rational obligation, then we may safely eliminate reference to obligation in these passages as inessential. For to say that we have a rational obligation to do X, is only to say that reason dictates or prescribes doing X. No limitation of right is involved.

Let us then turn to the interpretation of the phrases 'in foro interno' and 'in foro externo'. Here we accept the excellent analysis provided by Warrender, who shows that Hobbes has two meanings intertwined:

1. An obligation to 'endeavour peace' (*in foro interno*) is contrasted with an obligation to perform the specific external actions which the laws of nature apparently prescribe to each individual (*in foro externo*).
2. Obligation to a system of laws, interpreted and administered by the individual conscience (*in foro interno*), is contrasted with obligation to a system of laws laid down by an external human lawgiver and administered by an external human judge. (This contrast may be said to provide also, by analogy, a second meaning for the expression, obligation *in foro externo*, though Hobbes does not use these words in such a context.)[1]

The first pair of meanings is entirely independent of obligation, in Hobbes's usual sense, although it can be expressed in terms of rational obligation. Reason dictates that we endeavour to fulfil the laws of nature to the extent compatible with our security. Thus we seek peace in a condition of

---

[1] Howard Warrender, op. cit., pp. 70–1.

war, and we keep peace—i.e. perform the determinate actions necessary to the preservation of peace—in a condition of peace. The difference between 'in foro interno' and 'in foro externo' is the difference external conditions impose on the manner in which we apply the laws of nature. If the laws of nature are not considered as obliging, then the two manners in which we apply them do not represent different ways in which they oblige.

But the second pair of meanings is not independent of obligation. For the difference here concerns the manner in which the laws of nature are *enforced*, and thus must be related to them as laws, not as rational precepts.

We may deal with the second term first. 'Obligation to a system of laws . . .' is clearly obligation to civil law. And this obligation is based on our covenant to obey the civil sovereign. It raises no problems for our view that the laws of nature do not as such create obligations.

But what about 'obligation to a system of laws, interpreted and administered by the individual conscience'? These laws are not civil laws; how then do they come to oblige us? A later passage in *Leviathan* provides the answer:

> And the same law, that dictateth to men that have no civil government, what they ought to do, and what to avoid in regard of one another, dictateth the same to commonwealths, that is, to the consciences of sovereign princes and sovereign assemblies; there being no court of natural justice, but in the conscience only; where not man, but God reigneth; whose laws, such of them as oblige all mankind, in respect of God, as he is the author of nature, are *natural*; and in respect of the same God, as he is King of kings, are *laws*. (*E.W.* iii, pp. 342–3)

The laws of nature oblige in conscience as divine laws, the laws of the King of kings. In the absence of civil society the only way of interpreting the laws of nature as laws is to consider them divine laws, obliging as the commands of God, and enforced in the conscience of the individual, 'where not man, but God reigneth'. To say this is not to deny that in the state of nature the laws of nature are primarily rational precepts. That they also oblige as God's commands may reinforce their effectiveness, but it does not determine their role as conditions of preservation. And it is as conditions of preservation that they tell us, not oblige us, to oblige ourselves.

This completes our defence of the claim that the laws of nature do not create obligations. We have found it necessary to admit that Hobbes's secondary definition of these laws, as commands of God, obliging in virtue of our obligation to obey God, does play some part in his account of the laws of nature, particularly as the laws affect the sovereign. To this extent our claim is an overstatement. But we have not found it necessary to suppose that the laws of nature are more than rational precepts, not in themselves creating obligations, in order to explain the primary role of these laws, and to account for what Hobbes says about them, in that primary role. In these most important respects our claim is upheld.

C. *Obligation, Motivation, and Reason*

Hobbes finds his account of obligation and covenant confronted with one pervasive problem—a problem whose solution takes him from moral theory to political theory. At the practical level, which is of greatest interest to Hobbes, the problem is to ensure that men actually perform their covenants. Security is not attained merely by entering into covenants, but only by adherence to them. And this adherence does not follow automatically from the fact that men have given up certain rights and taken on certain obligations, for as Hobbes notes, the signs of renunciation or transfer of right 'are the BONDS, by which men are bound, and obliged: bonds, that have their strength, not from their own nature, for nothing is more easily broken than a man's word, but from fear of some evil consequence upon the rupture' (*E.W.* iii, p. 119). Hobbes's pessimistic conclusion is that 'covenants, without the sword, are but words, and of no strength to secure a man at all' (*E.W.* iii, p. 154). At this practical level, then, we may consider the problem to be one of *security*.

But we may formulate the problem also both at the material and the formal levels of our analysis of Hobbes's theory. At the material level the problem is to show that men have sufficient motivation to adhere to the covenants which they have sufficient motivation for undertaking. Just because a man has covenanted to perform some action, it does not follow that, when the time for performance comes, he will be motivated to perform the action. If he is so motivated, then the covenant will be fulfilled, and to this extent the problem of security will be met. If it can be shown that men are always motivated to fulfil the covenants they have been sufficiently motivated to undertake, then the problem of security will be fully met. At this level, then, we have the problem of *motivation*.

But motivation and reason are linked for Hobbes. Reason is instrumental to the attainment of man's chief end, preservation. And so at the formal level we may state the problem of *reason*: to show that men have sufficient reason to adhere to the covenants which they have sufficient reason to undertake. Hobbes presents the problem in this way in his discussion of 'the fool' (*E.W.* iii, pp. 132–3), which we quoted earlier. But we should note immediately that, important as a formal statement of the problem is in making clear the issues involved, formal analysis itself provides no solution.

With this caveat in mind, we proceed to develop the problem of reason. Suppose two persons, A and B, enter into a covenant. There are then four possible outcomes:

1. Both A and B keep the covenant;
2. Neither A nor B keeps the covenant;
3. A, but not B, keeps the covenant;
4. B, but not A, keeps the covenant.

We now adopt a few notational devices, whose use is sufficiently indicated in these instructions:

'R(A&B)e' is read 'it is reasonable for A and B to enter the covenant';
'R(A)k' is read 'it is reasonable for A to keep the covenant';
'R(A) (1, 2)' is read 'it is reasonable for A to choose outcome (1) rather than outcome (2)'.

We adopt, as a simplifying axiom, the proposition that the reasonableness of choosing outcomes is strongly ordered:

For any person P and any outcomes m and n, either R(P) m, n) or R(P) (n, m).

The problem of reason, then, is to show:

If R(A&B)e then R(A&B)k.

To make clear the difficulty in establishing this, we analyse the reasonableness of entering and keeping covenants in terms of the reasonableness of choosing one outcome rather than another. The first step in the analysis of R(A&B)e seems evident; it is

If R(A&B)e then R(A&B) (1, 2).

This proposition imposes the condition that the outcome of making and keeping the covenant be more reasonable for both than the outcome of making and not keeping it. If this were not so, then the covenant would be entirely pointless.

Next, consider these propositions:

If R(A)e then R(B) (2, 4).
If R(B)e then R(A) (2, 3).

The first of this pair imposes the condition that B find unilateral adherence less reasonable than mutual violation. If this were not so, then B would perform the action required by the covenant whether A did or not, and so A would have no reason to enter the covenant. The second member of the pair imposes a similar condition on the reasonableness of B entering the covenant.

These three propositions, we suggest, give an analysis of R(A&B)e. Thus we have:

R(A&B)e if and only if R(A&B) (1, 2) and R(A) (2, 3) and R(B) (2, 4).

The first step in the analysis of R(A&B)k is parallel to that for R(A&B)e:

If R(A&B)k then R(A&B) (1, 2).

The same reasoning establishes the condition for keeping covenants as for making covenants.

Now consider these propositions:

If R(A)k then R(B) (1, 3).
If R(B)k then R(A) (1, 4).

The first of this pair imposes the condition that B find mutual adherence more reasonable than unilateral violation. If this were not so, then B would find it reasonable not to adhere if A adhered. But then, in virtue of R(A) (2, 3), A would not find adherence reasonable. The second member of the pair imposes a similar condition on the reasonableness of B adhering to the covenant.

It might be objected that B may not have the option of choosing between outcomes (1) and (3), or A between outcomes (1) and (4), so that these conditions may be inoperative. This may indeed be so, and we shall have to consider the effects of such situations. But at this point in our analysis we assume that these options do exist. And so we propose for our analysis of R(A&B)k:

R(A&B)k if and only if R(A&B) (1, 2) and R(A) (1, 4) and R(B)(1, 3).

The problem of reason arises, then, because the conditions in the analysis of R(A&B)e may be satisfied without the last two conditions in the analysis of R(A&B)k being satisfied. If this is so, then reason dictates violation of the agreement, in any situation in which the actions of each party are independent of the actions of the other. We may show this by a simple matrix, in which the order of reasonableness of the various outcomes is represented by the numerals 1 to 4, reasonableness for A being indicated first.

		B	
		adheres	violates
A	adheres	2, 2	4, 1
	violates	1, 4	3, 3[1]

[1] Readers familiar with the theory of games will recognize this matrix as a variant of the Prisoner's Dilemma. The analysis undertaken in this section owes

Note first that the situation described in the matrix meets the conditions for R(A&B)e. Mutual adherence is more reasonable than mutual violation for both, and neither finds unilateral adherence more reasonable than mutual violation. But the conditions for R(A&B)k are not satisfied. Each finds unilateral violation more reasonable than mutual adherence.

Suppose A and B to be acquainted with this matrix. A notes that whatever B does, it is more reasonable for him to violate. If B adheres, A attains the most reasonable outcome by violating. If B violates, A avoids the least reasonable outcome by violating. Hence reason dictates violation to A, and similarly to B. And so the agreement is pointless, unless A and B can ensure that the adherence of each is a condition of adherence by the other. Although each has reason to enter into the agreement, each also has reason to violate it.

This rather abstract analysis may be illustrated by reference to a concrete problem of the present time—the problem of disarmament.[1] For the purposes of discussion we shall suppose that there are only two nations, and we shall suppose that an outcome is reasonable for a nation to the extent to which it is preferred by that nation.

Now we suppose that these nations agree that a mutual arms race is disadvantageous for each, involving both great expense and great danger. Hence the first condition for making and keeping a disarmament agreement is met; both agree that it would be preferable for each to keep such an agreement than for both to violate it, and return to the arms race. Furthermore, neither considers unilateral disarmament reasonable; each prefers the arms race to being at the mercy of the other. Hence the conditions for the reasonableness of making the agreement are met.

But each prefers unilateral disarmament by the other to

much to the theory of games, although no formal application of the theory to Hobbes is attempted. The reader will also find important similarities between this section and my article, 'Morality and Advantage', *Philosophical Review*, vol. lxxvi, no. 4, October 1967, pp. 460–75.

[1] This example is developed at some length in 'Morality and Advantage', pp. 464–8.

mutual disarmament. And so neither considers it reasonable to keep the agreement if the other does. Unless, then, each has some way of ensuring that the other keeps the agreement, the conditions for reasonableness of keeping are not met. And so a disarmament agreement between the two nations would be ineffective. And this, indeed, reflects the present state of disarmament negotiations. Although each nation would benefit from mutual disarmament, each knows that any agreement to disarm would be violated by others, and so would be pointless.

We may now return to Hobbes. The conditions for the reasonableness of making an agreement may be considered conditions for the validity of covenant. Our extended formal definition of obligation, it may be remembered, is that A has a natural obligation not to do X if and only if A has laid down the natural right to do X, in accordance with right reason. If entering a covenant is reasonable, then in making it A lays down a natural right in accordance with right reason and so assumes an obligation.

But if it is not reasonable for A to keep the covenant so made, then the problem of 'the fool' arises. For violation of covenant, which is contrary to obligation, is in accordance with reason. And this entails the collapse of the Hobbesian moral system.

How, then, does Hobbes attempt to meet the problem of reason, to show that if $R(A\&B)e$, then $R(A\&B)k$? Three possible attacks present themselves. The first is to show that A and B do not have the options of choice between outcomes which give rise to the second and third conditions in the analysis of $R(A\&B)k$. These conditions may then be dropped, and $R(A\&B)e$ will entail $R(A\&B)k$.

The second is to insist that these conditions are always satisfied, that it is never the case that, given $R(A\&B)e$, either $R(A) (4, 1)$ or $R(B) (3, 1)$. Mere logic will not suffice here, but an appeal to material considerations may demonstrate this.

The third possibility is to arrange for the imposition by *force majeure* of what may be considered either a change in the relative reasonableness of certain outcomes, or an elimination of certain options of choice between outcomes. Hobbes considers each of these possibilities in turn.

Our analysis has abstracted from the temporal dimension involved in making and keeping, or violating, covenants. If we consider this dimension, we realize that the acts A and B are required to do in order to keep their covenant need not be simultaneous, and if they are not, then one of the troublesome options is eliminated. For let A be the person who is to perform first. If he violates the covenant, then B may avoid outcome (4) by violating as well. Since R(B) (2, 4), B will find violation reasonable. Hence the condition R(A) (1, 4) may be dropped, since A cannot choose between outcomes (1) and (4).

Unfortunately, as Hobbes recognizes, this does not improve matters. For if A performs first, then B may avoid outcome (1) without risking outcome (4). B has an effective option to choose between outcomes (1) and (3), and if R(B) (3, 1), B will find it reasonable to choose (3). Thus Hobbes says:

> For he that performeth first, has no assurance the other will perform after; because the bonds of words are too weak to bridle men's ambition, avarice, anger, and other passions, without the fear of some coercive power; which in the condition of mere nature . . . cannot possibly be supposed. And therefore he which performeth first, does but betray himself to his enemy; contrary to the fight, he can never abandon, of defending his life, and means of living. (*E.W.* iii, pp. 124–5)

To make this even clearer, suppose that the situation is of the type illustrated by our matrix. Since A is to perform first, B may choose to adhere or to violate in the light of what A does. And since violation in these conditions is always more reasonable for B than adherence, B reasonably chooses to violate. If A is aware of B's situation, he can predict this choice prior to making his own decision, and so he too must find violation reasonable. The covenant is thus pointless.

Consideration of the temporal dimension, then, eliminates only one of the conditions which prevents R(A&B)e from entailing R(A&B)k. What is wanted is some way of eliminating both conditions. For then, given R(A&B) (1, 2), which must be granted if A and B are to have sufficient reason for entering into the covenant, each will choose to perform.

But Hobbes instead attacks the problem in a different way. In the passage quoted he does not actually commit himself to the possibility that R(B) (3, 1). Instead he refers to certain passions, which lead the second performer to violate his covenant. He says nothing about the reasonableness of the violation.

It may seem obvious that reason, as Hobbes conceives it, must at times dictate violation. For if we add material considerations to the purely formal ones we have considered thus far, we note that reason is taken by Hobbes to prescribe the means to man's ends, and chiefly man's chief end, preservation. Common sense may seem enough to convince us that sometimes a man stands to gain, in terms of his power to preserve himself, by violating his covenants if the other party has performed first. What Hobbes says seems to lend support to this view.

It is of itself manifest that the actions of men proceed from the will, and the will from hope and fear, insomuch as when they shall see a greater good or less evil likely to happen to them by the breach than observation of the laws, they will wittingly violate them. (*E.W.* ii, p. 63)

Therefore before the names of just, and unjust can have place, there must be some coercive power, to compel men equally to the performance of their covenants, by the terror of some punishment, greater than the benefit they expect by the breach of their covenant; . . . (*E.W.* iii, p. 131)

But these passages do not actually commit Hobbes to the view that breach of covenant is reasonable. He speaks of what men see, or of the benefit they expect, but he does not say that men see, or expect, what actually is the case. And in *Leviathan*, though not in Hobbes's earlier political writings, he proceeds to deny that men judge correctly in supposing that they will benefit from breach of covenant.

The first step is to set against those passions which lead men to violate their covenants, a passion which supports the covenants.

The passion to be reckoned upon, is fear; whereof there be two very general objects: one, the power of spirits invisible; the other, the power of those men they shall therein offend. Of these two, though the former be the greater power, yet the fear of the latter is commonly the greater fear. The fear of the former is in every man, his own religion: which hath place in the nature of man before civil society. The latter hath not so; at least not place enough, to keep men to their promises; . . . So that before the time of civil society, . . . there is nothing can strengthen a covenant of peace agreed on, against the temptations of avarice, ambition, lust, or other strong desire, but the fear of that invisible power, which they every one worship as God; . . . (*E.W.* iii, p. 129)

But this fear may not suffice. And so Hobbes is led to his ultimate argument. We have already quoted the long passage about 'the fool' (*E.W.* iii, pp. 132–3), in which Hobbes states the case for supposing that breach of covenant may be reasonable. We now quote the long passage in which he replies.

For the question is . . . where one of the parties has performed already; . . . whether it be against reason, that is, against the benefit of the other to perform, or not. And I say it is not against reason. For the manifestation whereof, we are to consider; first, that when a man doth a thing, which notwithstanding any thing can be foreseen, and reckoned on, tendeth to his own destruction, howsoever some accident which he could not expect, arriving may turn it to his benefit; yet such events do not make it reasonably or wisely done. Secondly, that in a condition of war, . . . there is no man who can hope by his own strength, or wit, to defend himself from destruction, without the help of confederates; . . . and therefore he which declares he thinks it reason to deceive those that help him, can in reason expect no other means of safety, than what can be had from his own single power. He therefore that breaketh his covenant, and consequently declareth that he thinks he may with reason do so, cannot be received into any society, that unite themselves for peace and defence, but by the error of them that receive him; nor when he is received, be retained in it, without seeing the danger of their error; which errors a man cannot reasonably reckon upon as the means of his security: and therefore if he be left, or cast out of society, he perisheth; and if he live in society, it is by the errors of other men, which he could not foresee, nor reckon upon; and consequently against the reason of his preservation; . . . (*E.W.* iii, pp. 133–4)

Hobbes's argument is, then, that violation of covenant cannot be *expected* to be advantageous, although it may actually *be* advantageous. Consequently a man cannot, with right reason, judge that violation of covenant will truly conduce to his preservation. And so violation cannot be more reasonable than adherence.

We have seen that R(A&B)e does not formally entail R(A&B)k. But Hobbes argues that if we take material considerations properly into account, we find that when R(A&B)e holds, then the conditions R(A) (1, 4) and R(B) (1, 3) must either both hold, or one holds and the other is irrelevant, because the person is unable to choose between the two outcomes. And so R(A&B)k holds.

Hobbes does not deny that, from a short-term point of view, a man may judge with reason that he should violate some of his covenants. Rather he insists that the long-term effects of such violation must always be expected to be sufficiently adverse to outweigh any short-term benefits. Thus we may say that although a situation in the short-run may be represented by a matrix of the type we have considered, in the long-run the correct representation (ignoring the complications introduced by discussion of the first performer) must be:

		B	
		adheres	violates
A	adheres	1, 1	4, 2
	violates	2, 4	3, 3

The situation described in this matrix meets the conditions for R(A&B)e. Mutual adherence is more reasonable than mutual violation for both, and neither finds unilateral adherence more reasonable than mutual violation. And the conditions for R(A&B)k are also satisfied, for each finds mutual adherence more reasonable than unilateral violation.

Now it is true that, if A and B are acquainted with this matrix, neither will find adherence the more reasonable course of action, whatever the other does. Violation might still be considered the safe strategy, in the sense that it ensures avoidance of the worst possible outcome for each—unilateral adherence. But since each stands to gain from adherence as long as the other adheres, the parties to a covenant in a situation of the type described by the matrix should have no difficulty in co-ordinating their choice of actions to bring about mutual adherence.

Hobbes has now provided a partial solution to the problem of motivation. Since adherence to covenants must always be expected to conduce to preservation, a man who reasons correctly, and who is not overwhelmed by occurrent passions, will find himself with a sufficient motive to keep his covenants. But errors in reasoning, and the force of particular passions, both lead men astray. A man may not, then, be actually motivated to fulfil his covenants.

Thus Hobbes has not yet solved the problem of security. In order to ensure that men are actually motivated to fulfil their covenants, he considers that there must be an earthly power sufficient to hold men to these covenants, and so he is led to the need for civil society in which there is a sovereign with the required power.

We may consider the effect of this earthly power in either of two ways. We may suppose that it effectively rules out the possibility of all outcomes save (1). Men must, then, keep their covenants; they have no choice in the matter. Or we may suppose that it imposes sufficient penalties for violation of covenant, that fear of these penalties is sufficient, without further deliberation, to lead men to opt for adherence, and so for outcome (1). The other outcomes will then be considered possible, but evidently unreasonable.

Let us sum up the course of our argument. Hobbes's problem is to ensure that men actually perform their covenants. Considered abstractly, as a problem about the reasonableness of adhering to covenants, we discovered that the fact that it was reasonable for men to enter a covenant did not of itself entail that it was reasonable for them to keep the covenant. But, we noted, Hobbes argues that if a covenant is made for the sake of preservation, then we must always expect preservation to be furthered by keeping the covenant, and so we do always have reason to keep it. But, although this provides a motive for keeping covenants, mistaken reasoning and strong occurrent passions may keep this motive from being operative. Thus the problem of security is solved only if there is a force sufficient to hold men to their covenants. And this takes us to the starting-point of Hobbes's political theory, which we shall examine in the next chapter.

Before leaving the present topic, however, I wish to propose a reformulation of the issues we have been discussing. The purpose of the reformulation is to provide an alternative to the rather implausible manner in which Hobbes prevents his moral theory from collapsing into inconsistency, without sacrificing either the substance of the theory, or affecting Hobbes's solution to the problem of security. The reformulation will, however, require certain modifications in Hobbes's view of obligation and covenant.

Hobbes's reply to 'the fool' is not very convincing. It seems simply false to maintain that a man can never expect breach of covenant to be conducive to his preservation. Furthermore, this reply is not necessary to Hobbes's solution of the problem of security. For if Hobbes were to admit that, in the state of nature, violation of covenant were sometimes reasonable, he could then introduce the sovereign, not only to hold men to their covenants against error and passion, but also to hold men to their covenants by imposing penalties sufficient to outweigh the *real* advantages of violation.

Hobbes needs his reply to 'the fool', only to avoid the possibility that a covenant might be reasonably made, and so valid and obliging, but not reasonably kept. It is at this point that my reformulation is to be introduced.

There is something odd about claiming that it is logically possible for it to be reasonable to enter into a covenant, yet not reasonable to keep that covenant. Our analysis of R(A&B)e and R(A&B)k has shown that in at least one sense this oddity is only superficial. Yet it might still be urged that it is misleading to speak, as we have done, about the reasonableness of entering a covenant, or to read 'R(A)e' as 'it is reasonable for A to enter the covenant'. Rather, it might be suggested, it is reasonable for A to enter into a covenant if and only if both the conditions present in the analysis of R(A)e, viz. R(A) (1, 2) and R(B) (2, 4), and the conditions present in the analysis of R(A)k, viz. R(A) (1, 2) and R(B) (1, 3), hold.[1]

To avoid confusion with what we have said previously, let us, rather than redefining reasonableness of entry into a covenant, instead introduce the notion of reasonableness of covenant. We adopt a new notational device, whose use is shown in this instruction:

'R(A)c' is read 'the covenant is reasonable for A'.

We define reasonableness of covenant by the equivalence:

$$R(A)c = R(A)e \text{ and } R(A)k.$$

We now say that a covenant is valid, and obliging, and breach of it unjust, if and only if the covenant is reasonable. Thus we interpret our formal definition of 'obligation' in such a way that a person is said to lay down a right only when both imposing such a restriction on himself, and adhering to that restriction, are in accordance with reason.

The formal problem raised by 'the fool' now vanishes. For if it is contrary to reason to keep one's covenant, then it is, in effect, no covenant, and one is not unjust in doing what it forbids. But the material problem raised by 'the fool' remains. For what he now must be taken to say is that sometimes a man may find it conducive to his preservation not to adhere to agreements which must be kept if men are to be secure. Or in other words, not all those covenants which are required for peace are reasonable.

The position taken by 'the fool' may now be granted—he is, after all, no fool. For his argument no longer undermines Hobbes's moral theory, but instead reinforces Hobbes's claim that 'covenants, without the sword, are but words, and of no strength to secure a man at all' (E.W. iii, p. 154).

---

[1] The analysis of R(A)e does indicate those positive conditions which must be met, if it is to be reasonable for A to enter into the covenant. Hence we might say: it is reasonable for A to enter into a covenant if and only if the conditions present in the analysis of R(A)e hold, unless (and only unless) the conditions present in the analysis of R(A)k do not hold.

Hobbes need no longer reply that if a covenant is undertaken to secure peace and preservation it must be reasonable to fulfil it. Instead he can admit that, in the state of nature, it may not be reasonable to fulfil such a covenant—and therefore some power is needed to *make* it reasonable to fulfil it.

On the view implicit in our reformulation, then, in the state of nature it may be the case that R(A&B)e, but not R(A&B)k and so not R(A&B)c. The role of the sovereign must be to ensure that whenever it is the case that R(A&B)e, R(A&B)k and so R(A&B)c also hold.

So far, it may be thought, our reformulation is quite unobjectionable from Hobbes's point of view. But it does, in fact, run counter to certain features of his doctrine of obligation. Hobbes never suggests that a man has an obligation to perform an action only if that action is to his advantage. He does insist, in reply to 'the fool', that obligation and true expectation of advantage do coincide, but he does not make it a condition of obligation that this coincidence obtains. Under our reformulation, however, this does become a condition of obligation; not only do we undertake obligations only for prudential reasons, but the undertaking extends only to what we have prudential reasons for carrying out.

As we suggested before, such a view 'would divest the claim that we had an obligation of its usual significance'. But can Hobbes consistently avoid it? This is one of the questions we must now face in assessing Hobbes's moral theory.

4. MORALITY OR PRUDENCE?

Hobbes's moral system is an extremely impressive edifice. Its foundations are strictly subjectivistic—good 'is the object of any man's appetite or desire' (*E.W.* iii, p. 41). But on these foundations Hobbes erects 'the science of what is *good*, and *evil*, in the conversation, and society of mankind' (*E.W.* iii, p. 146). All men agree that war is their enemy, and so 'all men agree on this, that peace is good, and therefore also the way, or means of peace, which . . . are . . . the laws of nature, are good' (ibid.).

Subjective prudence fails as a guide to action. If men exercise their unlimited natural right in accordance with their own judgement, they find themselves in a war of all against all. The laws of nature, appealing to the reason of all men, require men to limit their natural right, accepting certain obligations one to another. Thus Hobbes claims to erect an objective morality from the ruins of subjective prudence.

At no stage of his argument does Hobbes lose sight of the viewpoint of the individual. Although his moral system provides a common viewpoint for all men, yet each man must see what is common as the outcome of his own particular concern with his own preservation and well-being. Thus each new step in the argument is justified to each man in terms of his own good. For at each step the alternative is the same: consent— to the laws of nature, to restriction of natural right, to civil society, to unlimited sovereignty—or perish.

It is therefore misleading to claim, as Warrender does, that '(t)he laws of nature . . . are not maxims for personal success, nor even personal rules for keeping alive. . . . they are general principles for the preservation of humanity'.[1] For the alternative is wrongly put. The laws of nature are maxims for the preservation of the individual, but they hold for every individual. If each man preserves himself, then all men—society— is preserved. But the basic consideration is always the preservation of the individual.

Hobbes may with reason be regarded as the bourgeois, or individualist political philosopher, par excellence. His doctrine may seem strange to those who connect bourgeois attitudes with *laissez-faire*, and with the view that the best government is that which governs least. But Hobbes's argument is in many respects more consistent than that of many of his fellow individualists. For Hobbes takes seriously both the supposition that all individuals are equal, and the view, implicit in many bourgeois thinkers, that all men are naturally selfish. The resultant 'harmony of interests' is imposed, not by an invisible hand, but by the very visible hand of the absolute

[1] Howard Warrender, op. cit., p. 275.

sovereign. For all men agree that peace is good, but only a power imposed equally on all men can secure that peace.

But impressive as Hobbes's system is, we may question whether it is a moral system, in the sense in which we suppose that morality is not reducible to prudence. It is true that Hobbes's system transcends subjective prudence, but we may ask whether it is, or can be, more than common prudence.

Once again, we need not suppose that our question is one that would have been of great interest to Hobbes himself. As long as the laws of nature do their job, what matter if they are moral principles, or rules of common prudence? Indeed, an appeal to basic interest or advantage is surely the most effective appeal of all—and if it can be expressed in terms which have moral connotations, so much the better. Ideology is interest disguised as morality.

One may sympathize with the view that the effectiveness of the doctrine is more important than the label it carries. Yet the label is not unimportant if one supposes that an appeal to moral considerations, at least in some measure independent of interests, has its place even in the political sphere. For we may suspect that something distinctively moral is lacking from Hobbes's arguments. As we shall show, this suspicion is well founded. What is more, we shall trace what is lacking to Hobbes's psychology, thus refuting the view that his moral system can be treated as quite independent of his view of man.

What is lacking is not to be discovered by attention to the formal definitions of Hobbes's moral concepts. We have, indeed, found much to praise rather than to criticize, especially in Hobbes's conception of obligation. Thus Plamenatz confuses two quite distinct issues when he says, 'People who say that there is no place for obligation in the system of Hobbes want to deny that what he calls by that name is what we ordinarily understand by moral obligation. They want to say at least this much, and this much is surely true.'[1]

The question whether what Hobbes understands by the term 'obligation' is what we understand by it, is quite different

[1] John Plamenatz, op. cit., p. 75.

from the question whether there is place for moral obligation in Hobbes's system. As we have seen, Hobbes *understands* by obligation, self-imposed lack of right. This is not quite what we understand by it; we should query whether obligations must be self-imposed. But it is near enough, and it certainly does not rule out moral obligation.

On the other hand, the question whether there is place for moral obligation in Hobbes's system, is to be answered by considering whether the material conditions under which a man can actually be under obligation are such that we should classify his obligation as moral rather than prudential, or legal.

In insisting that Hobbes's moral concepts do not rule out the possibility of morality, we are not saying that our moral concepts are therefore identical with them. As we have noted, Hobbes's understanding of obligation is such that to be under an obligation (although not, it should be noted, to be naturally or rationally obliged), one must have put oneself under the obligation by renouncing some prior right. Many obligations may be explained by an appeal to this definition—obligations of promise-keeping, or obligations arising from membership in voluntary groups. In making a promise one renounces one's prior right not to do what one promises to do. In joining a club one renounces one's prior right not to do those things laid down by the club as 'Duties of Members'. But obligations arising from membership in non-voluntary groups, or obligations of gratitude, do not seem explicable on Hobbes's definition.

Hobbes would insist that we have no such obligations. There is no room in his system for a limitation on right which is not self-imposed. Our conceptions, then, differ from his— but not in a way which prevents his conceptions from being termed moral.

Another difference—perhaps merely terminological—arises from Hobbes's use of the phrase, law of nature. Perhaps what he means by law of nature is similar to what we mean by moral principle. In my moral vocabulary, at least, 'law of nature' has no place.

One important sense in which Hobbes's moral concepts are similar to our common ones is their connection with reason. To say that what it is right (or perhaps all right) to do is what one has (good) reason to do, is I think true and important. If the connection between reason and moral concepts is denied, as in the emotive theory of ethics, then an important conceptual revision is in fact being proposed. To say this is not to say that the emotive theory is wrong; perhaps our ordinary moral concepts are without foundation. But Hobbes in this respect defends our ordinary view.

Turning now from the formal to the material definitions of Hobbes's moral concepts, we note that both rights and obligations must have a prudential foundation. This is required by Hobbes's conception of reason, as instrumental to the ends of the individual, which are self-centred ends. Now this runs counter to our ordinary suppositions. I may undertake certain obligations—those, say, of a professor at a particular university—just because I judge it to be to my advantage to be a member of the faculty of that university. But I may equally undertake certain obligations—those, say, of a member of an organization supporting nuclear disarmament—because I suppose nuclear disarmament to be necessary to human well-being and survival.

But as we have already stated, the grounds on which we oblige ourselves do not themselves determine the character of our obligations. An obligation undertaken for prudential reasons is no less a moral obligation than one undertaken for moral reasons. If I borrow money from you, I have not only a legal but also a moral obligation to repay you, whether I borrow the money to contribute to the relief of starving children or to take a holiday in Mexico. Indeed, the moral obligation is, if anything, more stringent in the latter case than in the former.

It is not, then, Hobbes's prudential account of the grounds on which we undertake obligations which prevents us from classifying the obligations in his system as moral. We come, then, to the grounds for fulfilling our obligations. We admitted earlier that the absence of non-prudential reasons at this point would divest the claim that one has an obligation of its usual significance. Plamenatz puts the issue bluntly when he says, 'When someone is morally obliged, there is something he ought to do, whether it is to his advantage or not.'[1] This is what Hobbes is unable to say.

Or is he? Certainly if the reformulation of Hobbes's position advanced in the third part of this chapter were accepted, then to say that man has an obligation to do something would entail that it was to his advantage to do it. But we saw that Hobbes does not commit himself to this. His reply to 'the fool' may be put in a form to parallel Plamenatz's statement—'When someone is morally obliged, there is something he ought to do, which always as a matter of fact is to his advantage.' It does not follow from this that the obligation would lapse if the act were not to the person's advantage.

However, we may now confront Hobbes with a dilemma. Suppose a person, A, makes a covenant, and the conditions present in the analysis of R(A)e are satisfied—i.e. he considers mutual performance more reasonable for him than mutual non-performance, and (so that he has some incentive to covenant) he considers unilateral performance less reasonable for the other person than mutual non-performance. Now in practice it may always be advantageous to keep such a covenant. Hobbes holds this in his reply to 'the fool'. But it need not be so in theory. R(A)e does not entail R(A)k, so that it is in principle possible that unilateral non-performance is more reasonable for A than mutual performance.

Now, either A would have an obligation to keep the covenant, even if it were not to his real advantage to do so, or he would not. If he would, then he would have an obligation to do what is contrary to reason, and the Hobbesian moral system would collapse into theoretical inconsistency. If he would not, then having an obligation to keep a covenant entails that it is advantageous to keep it, and the Hobbesian 'moral' system would be purely prudential, without any place for moral obligation.

[1] John Plamenatz, op. cit., p. 75.

There is, however, a reply open to Hobbes. Let us suppose that, to avoid inconsistency, he takes the second alternative —i.e. accepts the reformulation suggested in the preceding section. Warrender holds that '(t)he provision that the agent must be able to have an adequate motive to perform what he is obliged to do, may be extracted from Hobbes's concept of moral obligation, and if this provision is denied, a different *type* of obligation is introduced'.[1] Taking his cue from Warrender, Hobbes may insist that it can make no sense to speak of someone having a moral obligation to do what he can have no adequate motive to do. Therefore, if one can never have an adequate motive to do what is to one's disadvantage, it cannot be the case that if someone is morally obliged there is something he ought to do, whether it is to his advantage or not.

The short reply to this is that if one can never have an adequate motive to do what is to one's disadvantage, then *moral* obligation is impossible. But this needs further explanation.

If a person, A, enters into an agreement with another person, B, then at least part of the force of our ordinary claim that A has a moral obligation to keep the agreement is given by:

$$\text{`R(A) (1, 2)' implies `R(A) (1, 4)'.}$$

If it is more reasonable for A that both fulfil the agreement than both violate it, it follows necessarily that it is more reasonable for A that both fulfil the agreement than that he unilaterally violate it.

The relation between 'R(A) (1, 2)' and 'R(A) (1, 4)' is not one of logical entailment. As we have seen, it is logically possible to hold the first and deny the second—this is Hobbes's problem of reason. But we take rejection of this implication to show moral unreasonableness. Why we do so, or whether we are justified in doing so, are questions which I am not here concerned to answer. What I am concerned to do is to set out a condition of moral reasonableness, which cannot be accepted by Hobbes.

[1] Howard Warrender, op. cit., p. 93.

To clarify just what is intended in this implication, let us take an example. Suppose A agrees to pay B a certain sum of money at the end of the month, and B agrees to repair the roof of A's house now. Then if it is more reasonable for A that B repair his roof and that A pay B the agreed sum than that B not repair his roof and A not pay him, it follows necessarily that it is more reasonable for A that B repair his roof and A pay B the agreed sum than that B repair his roof and A not pay him. If B repairs A's roof, and at the end of the month A refuses B payment, A is, in our view, a morally unreasonable man.

Note first that 'advantageous' cannot be substituted here for 'reasonable'. It may well be more advantageous for A to refuse B payment, and so have his roof repaired without cost, than to pay B. And this helps to clarify the substance of the objection that Hobbes cannot provide for moral obligation. What is reasonable and what is advantageous cannot be completely coincident.

Second, the implication does not require that any occurrence in some sense equivalent to outcome (4) be less reasonable for A than (1). Suppose B is a friend of A's, who is willing to repair the roof of A's house without charge. If B does this, then the outcome is materially the same for A as if B repairs the roof under the agreement but is not paid. But clearly it may be more reasonable for A to have his roof repaired without charge, than to pay for its repair. Outcome (4) is less reasonable than outcome (1) for A, only in so far as it involves A's violation of his agreement.

Finally, the implication does not require that any action which might lead to outcome (4) be less reasonable for A than fulfilling his agreement, and so bringing about outcome (1). Suppose that after B repairs A's roof, but before the end of the month, A's child becomes seriously ill and requires extensive medical treatment at great immediate expense to A. If A obtains this treatment for his child he will be unable to pay B at the end of the month, and so will unilaterally violate their agreement. It does not follow from the implication that securing this treatment for his child is less reasonable for A than adhering to his agreement with B, even though B has already fulfilled his part of the agreement. Considered in itself, unilateral violation by A is less reasonable for him than mutual adherence, but unilateral violation may be part of a course of action more reasonable than mutual adherence.

In Hobbes's system the proposition ' "R(A)(1, 2)" implies "R(A) (1, 4)" ' does not hold. A Hobbesian man must always determine the rationality of keeping an agreement rather than unilaterally violating it by considering which course of action is more conducive to his preservation. Thus whether or not 'R(A) (1, 4)' holds has no necessary connection with whether or not 'R(A) (1, 2)' holds. Thus Hobbes's system has no place for moral obligation. The conditions under which a man has an obligation are not those which we classify as moral, rather than merely prudential.

This conclusion has quite general application. In no system of rational prudence, in which all reasons for acting must reduce to considerations of what, in each situation, is most advantageous for the agent, can moral obligation be introduced. Furthermore, as I have argued elsewhere, in any such system certain benefits are unobtainable which may be obtained in a genuinely moral system.[1] Rationally prudent men, oddly enough, cannot secure certain benefits which are open to moral men.

The proof is quite simple. Consider once again our original matrix, which we interpret now as indicating the relative *advantage* to the agents, of the outcomes of a possible agreement:

		B	
		adheres	violates
A	adheres	2, 2	4, 1
	violates	1, 4	3, 3

Suppose that the relative advantage of no agreement is equal to that of mutual violation. Then, if A and B are rationally

---

[1] Cf. 'Morality and Advantage'.

prudent men, they cannot reap the benefits of such an agreement. For since prudence will always require each to violate
the agreement, it will be pointless for them to make it.

In practice it may be possible to overcome this problem by
providing some means of penalizing violation, so that it is no
longer relatively more advantageous than adherence. And this
brings us back to the Hobbesian sovereign, who is—for Hobbes
—the necessary means.

We have shown, then, that the Hobbesian 'moral' system is
nothing more than a system of common, or universal, prudence.
But to show this, we have had to appeal to Hobbes's account
of human nature, as manifested in the material definitions of
his 'moral' concepts. It is only the fact that men are necessarily bent on their own preservation, or more generally on
their own advantage, that prevents us from classifying
Hobbes's system as moral. In this way his psychology is
not only relevant to, but destructive of, his ethics.[1]

[1] It might be objected that our conclusion that Hobbes has no moral system
shows the inadequacy of our interpretation of that system. But our conclusion
depends solely on Hobbes's psychology. No alternative interpretation will alter
the fact that, given Hobbes's psychology, no truly moral system is possible.

Warrender does not seem to be clear about this. He distinguishes grounds of
obligation from validating conditions of obligation (op. cit., pp. 14–17), and then
interprets Hobbes in such a way that the grounds are independent of motivation. Motivation enters as a validating condition, in the form 'The individual
cannot be obliged to do that for which he cannot have a sufficient motive'
(pp. 24–5). He then takes Hobbes's claim 'that a person cannot have a sufficient
motive to kill himself' (p. 25) to be a psychological generalization required only
to apply Hobbes's concept of obligation. Thus he says, 'Hobbes says so much
about self-preservation that it is easily regarded as being central in his theory of
obligation. This is so far from being the case that it is not a part of that theory
as such, but an empirical postulate employed in its application' (p. 93).

If we take Hobbes's theory to be just the formal definitions of moral concepts
and their consequences, then of course Warrender is right. But then the theory
is neutral between morality and prudence. To determine whether Hobbes has
a moral system, we must appeal to the empirical postulate. Warrender shows no
indication of recognizing this.

III

INITIAL THEORY OF SOVEREIGNTY

THE title of this chapter indicates a departure from the usual interpretation of Hobbes's political theory. Although scholars have noted that there are differences between the presentation of the theory in *Leviathan* and the presentation in Hobbes's earlier political writings, little significance has been attached to these differences. There is a plausible explanation for this—the differences do not affect the material content of the theory. We wish to argue, however, that the differences amount to a major alteration in the formal structure of Hobbes's political theory—an alteration which adds both plausibility and interest to Hobbes's view of political society, but which Hobbes is unable properly to exploit because of the inadequacies in his psychology.

In this chapter, then, we shall examine Hobbes's initial theory, concentrating attention upon those weaknesses which are at least partially overcome by the alterations in *Leviathan*. We shall include a discussion of what Hobbes terms *sovereignty by acquisition*, since this is adequately explained only by the initial theory, and fits uncomfortably into the framework of the later system. Although we shall not forget the distinction we have made between formal and material considerations, we shall not pursue it in this chapter because Hobbes does not develop important and distinctive *political* concepts in his initial theory.

Before turning to the details, however, it will be well to establish one crucial distinction which Hobbes never sufficiently develops in any presentation of his political theory. This is the distinction between *sovereign right* and *sovereign power*.

Sovereign right is not merely the right to rule over other men. Every man has this right by nature (*E.W.* iii, p. 346). But sovereign right must be exclusive; in a given political body, there can be only one such right. And it must be more than permissive; the right of the sovereign must be correlative with the obligation of the subjects. For there can be no sovereign without subjects, and no subjects without obligation to obey.

Now Hobbes's political theory is primarily concerned with this sovereign right. He must determine how it is established and how far it extends. But parallel to these questions are questions about sovereign *power*. How is a supreme power established which enables its holder to enforce obedience? And how far does this power extend—how far can its holder compel obedience?

These questions about right and power are interdependent. For in Hobbes's view there can be no sovereign right without power. 'The obligation of subjects to the sovereign, is understood to last as long, and no longer, than the power lasteth, by which he is able to protect them' (*E.W.* iii, p. 208). Without this power, sovereign right would be no more than the natural right of every man to rule over the rest—a right which is without effect.

Conversely, there can be no *supreme* power without right. For supreme power is the power of life and death, and 'every man is supposed to promise obedience, to him, in whose power it is to save, or destroy him' (*E.W.* iii, p. 188). Supreme power enables its possessor to establish an effective right to rule, and so to acquire sovereignty.

But although right and power are for Hobbes intimately connected, they must not be confused. In showing how sovereign right is established, we must not fail to show that power accompanies this right, to make it effective. And in showing how power makes itself supreme, we must not fail to explain why this power can acquire right for itself. Hobbes provides an account of the latter, which indeed the quotation in the preceding paragraph already suggests. But he is confused about the former, and as we shall see, tends to move back and forth between the establishment of right and the establishment of power in a way which obscures the structure of his argument.

1. SOVEREIGNTY IN *DE CORPORE POLITICO* AND *DE CIVE*

Hobbes begins his account of the institution of common-wealth by speaking as if all that has gone before is of no avail. If men 'see a greater good or less evil likely to happen to them by the breach than observation' of the laws of nature, 'they will wittingly violate them' (*E.W.* ii, p. 63). And so each man must seek to guard himself against such breaches, and must retain the 'same primitive right of self-defence' (*E.W.* ii, p. 64) which is 'a right to all things'. Although men recognize that to secure peace they must give up this right to all things, in the state of nature they cannot safely do so.

Since awareness of the laws of nature is not a sufficient condition of peace, Hobbes considers what is needed so that men may be able to act on these laws. The first requirement is numbers. If a man is to preserve himself against his enemies he requires confederates, and a body of confederates large enough to deter any potential enemy from attack. The larger the groups, the less certain it is which would prove victorious in struggle, and so the less likely struggle becomes.

But confederates, however many, will not suffice unless they are agreed upon a common plan of action directed to mutual survival and well-being. This agreement, however, is dubious and precarious in a mere aggregate of men. For even if present danger enables them to agree on immediate defence, they have no regular procedures to maintain that agreement, and no way to avoid the divisive effects of the desire of each man to be foremost. Enduring consent requires *union*.

The formal definition of union, which is the key concept in Hobbes's earlier political writings, is 'the involving, or inclu-ding the wills of many in the will of one man, or in the will of the greatest part of any one number of men' (*E.W.* iv, p. 121). In union, one will determines the common plan of action. Thus agreement, once reached, is maintained, since a pro-cedure is now available for determining what the entire group should do, not only for the moment, but for the indeterminate future. And if this will has power commensurate with its scope, the desire of each man to be foremost can be held in check, reinéd to the service of the common good as determined by the one will.

Hobbes's account of the making of union changes between *De Corpore Politico* and *De Cive*. Because the change has important consequences for the structure of the argument, we must examine the relevant passages. In *De Corpore Politico* Hobbes argues:

> The making of union consisteth in this, that every man by covenant oblige himself to some one and the same man, or to some one and the same council, by them all named and determined, to do those actions, which the said man or council shall command them to do, and to do no action, which he or they shall forbid, or command them not to do. . . . And though the will of man being not voluntary, but the beginning of voluntary actions, is not subject to deliberation and covenant; yet when a man covenanteth to subject his will to the command of another, he obligeth himself to this, that he resign his strength and means to him, whom he covenanteth to obey. . . . And because it is impossible for any man really to transfer his strength to another, or for that other to receive it; it is to be understood, that to transfer a man's power and strength, is no more but to lay by, or relinquish his own right of resisting him to whom he so transferreth it. (*E.W.* iv, pp. 121–3)

In sum, every man obliges himself *to some man or council*—the prospective sovereign—to do what he commands and not what he forbids—that is, to resign to him his strength and means, which is, to relinquish his right to resist him.

In *De Cive*, Hobbes argues somewhat differently:

> This submission of the wills of all those men to the will of one man or one council, is then made, when each one of them obligeth himself by contract to every one of the rest, not to resist the will of that one man or council, to which he hath submitted himself; that is, that he refuse him not the use of his wealth and strength against any others whatsoever; for he is supposed still to retain a right of defending himself against violence: and this is called *union*. . . .

> But though the will itself be not voluntary, but only the beginning of voluntary actions; (for we will not to will, but to act); and therefore falls least of all under deliberation and compact; yet he who submits his will to the will of another, conveys to that other the right of his

strength and faculties . . . which to have done, because no man can transfer his power in a natural manner, is nothing else than to have parted with his right of resisting. (*E.W.* ii, pp. 68–70)

In sum, every man obliges himself *to every other man* not to resist the will of some one man or council—that is, to resign to him the use of his wealth or strength, which is to part with his right to resist him.

These accounts raise two problems, the first of which concerns the difference between them. To whom is one obliged? To the sovereign, as in *De Corpore Politico*, or to one's fellow covenanters, as in *De Cive*?

Technically, the latter is the only satisfactory answer. For the covenant or contract by which the sovereign is instituted is between the prospective subjects, and not between these subjects and the prospective sovereign. The sovereign is not a party to the covenant and so the subjects do not oblige themselves to him.

But Hobbes does want to insist that we are obliged to the sovereign as well as to our fellow covenanters. Having said this directly in *De Corpore Politico*, he does not consider further how this is possible. But in *De Cive*, recognizing that the covenant is between only the subjects, he does seek to show that they also have an obligation to the sovereign.

If there is no obligation to the sovereign, then the subjects can in theory mutually release themselves from the covenant establishing his right and power. Hobbes does not suppose such release could really occur, since it would require that everyone release everyone else—any one man dissenting, the covenant would stand. But to introduce a practical consideration seems beside the point, for the institution of sovereignty requires that everyone covenant with everyone else, and this is hardly a practical possibility. If release is theoretically possible, then the subjects can in principle deprive the sovereign of his position, and this Hobbes will not grant.

The argument takes this form:

For each citizen compacting with his fellow, says thus: *I convey my right on this party, upon condition that you pass yours to the same*: by which means, that right which every man had before to use his faculties to his own advantage, is now wholly translated on some certain man or council for the common benefit. Wherefore what by the mutual contracts each one hath made with the other, what by the donation of right which every man is bound to ratify to him that commands, the government is upheld by a double obligation from the citizens; first, that which is due to their fellow-citizens; next, that which they owe to their prince. (*E.W.* ii, pp. 91–2)

But what obligation to the sovereign does this establish? Each man agrees with each of his fellows to convey his right of nature to the sovereign. If each man then releases each of his fellows from this agreement, what obligation remains? Hobbes has insisted that 'an injury can be done to no man but him with whom we enter covenant, or to whom somewhat is made over by deed of gift, or to whom somewhat is promised by way of bargain' (*E.W.* ii, pp. 31–2). But the subjects make no covenant with the sovereign, they make no deed of gift to him, they promise him nothing by way of bargain. For in each of these cases the sovereign would have to signify acceptance —enter into the covenant, receive the deed, accept the bargain. And he does none of these.

As Hobbes points out, 'damaging and injuring are often disjoined. For if a master command his servant, who hath promised to obey him, to pay a sum of money, or carry some present to a third man; the servant, if he do it not, hath indeed damaged this third party, but he injured his master only' (*E.W.* ii, p. 32). Similarly, if two men covenant to obey a third person, and one does it not, he has damaged the third person, but has injured only the person with whom he covenanted. And if this person has released him from the covenant, then he has injured no one. The sovereign would be damaged if each man released each other from the covenant instituting him, but he would not be injured, and so the subjects have no obligation to him.

This problem is one to which we shall return in examining the argument of *Leviathan*. We shall find that although Hobbes still fails in his own attempt to account for an obligation to the sovereign, he does provide the premises from which an argument sufficient for his purposes might be derived. Here we can only leave the matter unresolved.

The second problem raised by these accounts of the institution of sovereignty concerns the relation of right and power. What precisely do the prospective subjects transfer or convey to the prospective sovereign? Do they transfer right— i.e. renounce their natural right in favour of the natural right of the sovereign? Or power? Or both?

Hobbes's account clearly alternates between a transfer of right and a transfer of power. There are differences in the two passages, and the confusion is more evident in *De Corpore Politico*. There, Hobbes passes from the supposition that each man obliges himself to obedience—that is, renounces his right to decide for himself what to do, to the supposition that each man resigns his strength—that is, his power, to the supposition that each man renounces his right to resist the sovereign—which seems to mean, not that each man obliges himself to obedience, but to non-hindrance.

In *De Cive*, Hobbes begins by supposing that each man obliges himself not to resist the sovereign—that is, renounces his right to resist, interprets this as resigning to the sovereign the use of his wealth and strength, takes this to mean that each man renounces the right to use his strength in favour of the sovereign, and concludes, as he began, that each man parts with his right of resistance.

The problem is evident. Hobbes imposes two opposed conditions on the act whereby the subjects institute the sovereign. On the one hand, he insists that the will of the subject is not voluntary; on the other hand, he insists that power is not literally transferable. From the first condition it follows that renunciation merely of right is ineffectual, since as long as the subject retains the power to do as he pleases he cannot but will to do so. From the second condition it follows that renunciation of power is impossible. From the two conditions taken together, it follows that the prospective subjects can only renounce their right and not their power, and that such renunciation is insufficient to institute the sovereign.

This is an unpalatable conclusion, and we must ask if Hobbes can avoid it. In part he can, but to show this we must examine more closely the steps in the institution of a commonwealth.[1]

The first step is the creation of a single will, a single centre of decision for the society. As we have seen, Hobbes supposes this to be effected by means of an agreement of every man with every man. The content of this agreement is that the wills of the majority must be taken as the will of everyone. In this way a democracy is created—a democracy being a civil society in which sovereign power is exercised by the whole body of citizens in assembly.

The second step is the establishment of a power of coercion, so that men may be held to the agreement made in instituting the sovereign. This power, which enables the sovereign to punish offenders, is termed the *sword of justice*. It arises from the act in which every man gives up his right to resist the sovereign—or more positively, 'when every man contracts not to assist him who is to be punished' (*E.W.* ii, p. 75). Hobbes observes that 'these kind of contracts men observe well enough, for the most part, till either themselves or their near friends are to suffer' (ibid.).

Each man recognizes an interest in maintaining society, and recognizes that the sword of justice is required to maintain society. Thus in each particular situation, each man will be disposed to favour the sovereign, unless his personal interests are adversely affected. This means that the sovereign will always have numbers on his side, at least as long as he seeks to punish only a few offenders on any one occasion. And in this way he can exercise a power greater than that possessed by any one subject, or any small group of subjects.

In effect, then, the sovereign's natural power is enhanced, because men recognize the necessity of a focal point for decision, and he is that focal point. To this extent the sword of justice can be established, and the problem of transferring

---

[1] The discussion following is based on *De Corpore Politico*, Second Part, chapter i, and *De Cive*, chapter vi.

power as well as right to the sovereign is resolved. But note that the resolution is at best partial, although the skilful sovereign will be able steadily to increase the actual power available to him, by using his position to convince more and more men of the advantage in giving him their active support, and so in the course of time may become truly supreme. Hobbes, of course, wished for absolute and unconditional supremacy in the sovereign, but this wish seems impossible of fulfilment.

The third essential step in instituting a body politic is the establishment of a power to defend the members against those without the society. This power, which like the sword of justice must be put in the hands of the sovereign, Hobbes terms the *sword of war*. The sovereign has 'the right to arm, to gather together, to unite so many citizens, in all dangers and on all occasions, as shall be needful for common defence against the certain number and strength of the enemy' (*E.W.* ii, p. 76).

But has the sovereign the power to exercise this sword? In time of war the sovereign must compel a very large body of his subjects, and indeed that body whose members have the greatest natural strength, to endanger their lives. Actual sovereigns exercise this power, but actual subjects are not Hobbesian men, bent on self-preservation. That Hobbes recognized this problem is sufficiently shown by the *Review, and Conclusion* of *Leviathan*, in which he adds to the laws of nature, '*that every man is bound by nature, as much as in him lieth, to protect in war the authority, by which he is himself protected in time of peace*' (*E.W.* iii, p. 703). The introduction of this law suggests that Hobbes was aware that the sovereign could not compel acceptance of his authority in war.

The conclusion suggested by our discussion is that the institution of sovereignty may be sufficiently effective in guaranteeing men against others of their society, but not in guaranteeing them against an external enemy. The sovereign can solve the problem of internal security, though not of external security. But such a conclusion is surely all that we can expect. It is consistent both with Hobbes's claim that civil society provides man with an escape from the state of nature, and with the evident fact that no society provides man with absolute security.

We have outlined the essential steps in the establishment of sovereignty. The basic form of civil society, established directly by covenant, is democracy, and the basic right of the sovereign is the right to exercise the two swords, of justice and war. There are, however, certain derivative considerations which need mention.

Among the corollary rights which Hobbes discusses, the right to establish and determine exclusive property-rights is of particular importance. In the state of nature every man has right to every thing; in civil society every man has exclusive right to some things. These rights are determined by the civil laws—the sovereign being, of course, the law-maker.

Since the sovereign is the source of property-rights, he is able to override them. My exclusive right to certain possessions extends only against my fellow subjects, and not against the government. Hobbes has no common ground with those who consider private property a pre-social or inalienable right; in this respect, at least, he cannot be considered a typical political theorist of 'possessive individualism'.[1]

Although democracy is basic in Hobbes's initial theory of sovereignty (cf. *E.W.* iv, pp. 138–9), monarchy is taken by Hobbes as both the typical and the best form of government. Monarchy—the rule of one, or aristocracy—the rule of a few, may result from democracy—the rule of all, if the sovereign assembly transfers its rights to one man or to a smaller group of men. Sovereign right is alienable.

It is also terminable. The sovereign may resign his position, although he is enjoined not to do so by the law of nature, since his resignation would restore the state of nature. Sovereignty over particular subjects may be terminated by

---

[1] The phrase is taken from C. B. Macpherson, *The Political Theory of Possessive Individualism: Hobbes to Locke*, Oxford, 1962, which attempts to treat Hobbes as a political philosopher of the possessive individualist persuasion.

direct release of those subjects from allegiance, as by banishment. If a monarch dies without a known heir, the sovereignty of necessity is terminated. And if the sovereign is overcome by enemies, so that he can no longer protect his subjects, then the sovereignty is terminated, and the subjects must seek protection either by accepting the victorious enemy as sovereign, or by instituting some new sovereign capable of defending them.

Sovereignty cannot, however, be forfeited. As long as the sovereign remains capable of protecting his subjects he cannot justly be deposed, nor can allegiance to him justly be renounced. Although Hobbes wants to insist that the sovereign can be injured by the disobedience of his subjects, he holds firmly to the view that they cannot be injured by his iniquity. The sovereign receives his position as a consequence of an agreement among the prospective subjects; he himself makes no agreements and accepts no conditions. Thus in dealing with his subjects he retains the full right of nature, the right of every man to rule over the rest, made effective by his subjects' renunciation of right, and obliging by their covenant not to resist him. Since the right of nature is unlimited, sovereign right is also unlimited.

Lest a false impression be created, it is essential to note that Hobbes considers it the duty of the sovereign to leave his subjects as free as the good of the commonwealth permits (cf. *E.W.* iv, p. 215; *E.W.* ii, pp. 178–9). Hobbes takes pains, not always acknowledged by his critics, to make clear what the law of nature—conceived here both as prudential reason and as the law of God—enjoins the sovereign to do. But the subjects cannot hold the sovereign to his duty; it is no injury to them if he violates the law of nature.

It is, however, an injury to God. And since the sovereign will is to be taken for the will of every man, it might be supposed that if the sovereign violates the law of nature, every man becomes guilty of wrongdoing against God. But Hobbes insists that only those whose natural wills are directly involved in the violation—the individual sovereign, or those members of the sovereign assembly who supported the violation—are so guilty. 'For a body politic, as it is a fictitious body, so are the faculties and will thereof fictitious also. But to make a particular man unjust, which consisteth of a body and soul natural, there is required a natural and very will' (*E.W.* iv, pp. 140–1).

Indeed, it would be unjust for the subjects not to obey their sovereign, even if his commands are against the law of nature. For they have covenanted not to resist him, and so to accept his judgement of what is to be done, of what is right and wrong. There is no law to which the subject can appeal against the civil law of the sovereign—no law which would justify the Nuremberg trials, or the trial of Adolf Eichmann.

The theory of sovereignty presented in Hobbes's earlier political works provides a very simple picture of the relation between sovereign and subject. This can be seen best by considering a monarchy in which the sovereign is an individual person. Every member of the society save one—the monarch—parts with his right of nature, in so far as that right would entitle him to resist or oppose the monarch. The monarch is left in full possession of the right of nature, but because it is not opposed by the corresponding rights of others it is an effective right. In this way natural right becomes sovereign right, not by its own augmentation, but by the diminution of the rights of others.

This simple and clear account does not wear well. In *De Corpore Politico* Hobbes avoids complicating it, but in *De Cive* additional features make their appearance. And these lead Hobbes towards the doctrine of *Leviathan*, in which this analysis of the relation between sovereign and subject is replaced by one dependent on the concept of authorization.

Hobbes introduces an important distinction in *De Cive*, when he argues, 'For it is one thing if I say, *I give you right to command what you will*; another, if I say, *I will do whatsoever you command*' (*E.W.* ii, p. 82). The clear implication of this statement, in its context, is that the subject does covenant the former but not the latter, and so gives the sovereign the right to command but is not thereby under obligation to obey.

But how can the subject give the sovereign the right to command what he will? The sovereign already possesses this right in possessing the right of nature. What is required is to make this right effective, and this entails that the subject renounce his right to resist—and so, since disobedience is resistance to the sovereign's command, that the subject oblige himself to obey. 'There is so much obedience joined to this absolute right of the chief ruler, as is necessarily required for the government of the city, that is to say, so much as that right of his may not be granted in vain' (ibid.).

If it be urged that Hobbes restricts the application of this distinction to those cases in which the subject is entitled to refuse to obey, because obedience would entail death (or, as Hobbes inconsistently suggests, what is worse than death), the argument is still incoherent. For it would then imply that the subject gives the sovereign the right to command his death, retaining only the right to disobey the command to kill himself or to allow himself to be killed.[1] But the subject cannot give the sovereign the right to command his death because he possesses no such right, not because he lacks it but because, as the material definition of the right of nature makes clear, the subject's right arises only within the framework of the means of preservation. Nor need he give the sovereign such a right, because the sovereign already possesses, by the right of nature, the right to kill whomsoever he pleases.

Hobbes does not avoid this incoherence in *Leviathan*. But there we shall be able to explain it. At this point we can note only that Hobbes's thought is moving away from the simple doctrine that the subject merely resigns his right to resist the sovereign, towards the view that the subject actually assigns to the sovereign the use of his right, in a manner inexplicable within the conceptual framework of *De Cive*.

That this is the direction of Hobbes's thought is made clearer in a passage following shortly on the one we have been

---

[1] That this restriction is intended by Hobbes is clear from the corresponding passage in *Leviathan*, *E.W.* iii, p. 204.

discussing. There Hobbes says: 'Neither can the city be obliged to her citizen; because, if he will, he can free her from her obligation; and he will, as oft as she wills; for the will of every citizen is in all things comprehended in the will of the city; . . .' (*E.W.* ii, p. 83). The sovereign exercises the will of the subjects; they have not merely renounced their right to will, but somehow transferred it to him.

In *De Corpore Politico* and *De Cive* Hobbes establishes only a negative relationship between sovereign and subject. The subject parts with his right to resist the sovereign; the sovereign is thereby enabled to make his natural right effective. But Hobbes wants to involve the subject more positively *in* the sovereign. He recognizes that society is a real union, a union expressed in the person of the sovereign, which contains that of each member of the society. And this requires that the will of each subject be expressed in the sovereign will, not merely that each subject renounce his own will in favour of the sovereign. Hobbes wants the sovereign to possess not just a natural right made effective by lack of opposition but the positive right of each of his subjects. Yet he has maintained that all transfer of right is really mere renunciation of right. Thus Hobbes seems to have made his task impossible from the outset, given the basic definitions of his moral theory. But as we shall see, a new conceptual departure will enable him to overcome this apparent impossibility, and model society on a positive relationship between sovereign and subject.

2. SOVEREIGNTY BY NATURE OR ACQUISITION

We have examined the method by which men who find themselves equally insecure in the state of nature erect a commonwealth to provide mutual security. But suppose some one man were able to subdue his fellows, each in turn, and make them subservient to him. Or suppose some men were naturally dependent on others for their survival—as children are dependent on their parents. These possibilities provide a second method for erecting a civil society—a method which, however, always plays a secondary role in Hobbes's thought about society. The two methods are clearly distinguished in this passage from *De Corpore Politico*:

> The cause in general, which moveth a man to become subject to another, is . . . the fear of not otherwise preserving himself. And a man may subject himself to him that invadeth, or may invade him, for fear of him; or men may join amongst themselves, to subject themselves to such as they shall agree upon for fear of others. And when many men subject themselves the former way, there ariseth thence a body politic, as it were naturally. From whence proceedeth *dominion, paternal* and *despotic*. And when they subject themselves the other way, by mutual agreement amongst many, the body politic they make, is for the most part, called a *commonwealth*, in distinction from the former, though the name be the general name for them both. (*E.W.* iv, pp. 123–4)

Note the primary role which Hobbes assigns to fear. The two methods for erecting civil society differ primarily in the object of the motivating fear—sovereignty by institution arises from the mutual fears prospective subjects have one of another, sovereignty by acquisition arises from the fear prospective subjects have of the prospective sovereign.

Because, as we said earlier in this chapter, sovereignty by nature or acquisition is best explained in terms of Hobbes's initial theory of sovereignty, we shall discuss it here. But we shall base the discussion, not so much on the earlier political writings but on *Leviathan*, which offers the clearest and most developed presentation of sovereignty by nature.

Despotic dominion arises when one man is vanquished, or expects to be vanquished, by another. But this situation is not formally sufficient to produce dominion. 'It is not therefore the victory, that giveth the right of dominion over the vanquished . . . . Nor is he obliged because he is conquered; that is to say, beaten, and taken, or put to flight . . .' (*E.W.* iii, p. 189).

Despotic dominion, like an instituted commonwealth, rests on covenant. Despotic 'dominion is then acquired to the victor, when the vanquished, to avoid the present stroke of death, covenanteth either in express words, or by other sufficient signs of the will, that so long as his life, and the liberty of his body is allowed him, the victor shall have the use thereof, at his pleasure' (*E.W.* iii, p. 189). Given Hobbes's view of man, the vanquished must offer such a covenant to the victor as the only way of preserving himself. Thus we may say that although, formally, sovereignty always depends on covenant, materially, it arises out of conquest.

The victor need not accept the covenant offered him. He may prefer either to kill or to enslave the vanquished. But if he does accept it, we may ask to what he thereby commits himself. For there is an evident problem here. A covenant is a *mutual* transfer of right. But if the victor transfers some right to the vanquished, he will have some obligation to him, and this is expressly contrary to Hobbes's insistence that the sovereign is not obliged to his subjects. Indeed, for the victor to accept the covenant is to make himself a party to it, and this is contrary to Hobbes's doctrine that civil society is not founded on a covenant between sovereign and subject—that the sovereign is not a party to the political covenant.

Hobbes conveniently avoids taking notice of this problem. But it fortunately proves to be more apparent than real. For in fact the victor *commits* himself to nothing in accepting the covenant of the vanquished. The vanquished promises obedience, *so long as* life and corporal liberty is allowed him. The victor, then, accepts the covenant merely by allowing the vanquished life and corporal liberty. He does not, and must not, covenant to allow life and liberty to the vanquished; to do this would deprive him of the right of killing or enslaving the vanquished at his pleasure. And this would be an obligation of the sovereign to the subject, contrary to Hobbes's basic doctrine of sovereignty. As Hobbes expressly says, 'nor is the victor obliged by an enemy's rendering himself, without promise of life, to spare him for this his yielding to discretion; which obliges not the victor longer, than in his own discretion he shall think fit' (ibid.) And to be obliged at one's own discretion is not to be obliged at all; 'For he is free, that can be free when he will: nor is it possible for any person to be bound to himself; because he that can bind, can release; and therefore he that is bound to himself only, is not bound' (*E.W.* iii, p. 252).

All the victor does, then, in accepting the covenant, is—literally or metaphorically—to take away the sword from his enemy's throat. The vanquished is then obliged, by his offer of covenant, to obey the victor. If the victor at some later time chooses to kill or to enslave the vanquished the covenant is terminated and the obligation to obedience ceases. But no man is obliged to obey someone who seeks his death, and Hobbes insists that a slave has no obligation to his owner. And of course the dictates of nature militate against the victor terminating the covenant, unless the vanquished fails to perform the promised obedience.

The covenant which establishes despotic sovereignty is technically degenerate, since it does not involve a mutual transfer of right. Rather it is a promise made on certain conditions. Thus we may say that Hobbes's account of despotic sovereignty involves a formal inconsistency, the following three propositions being jointly incompatible:

(i) Despotic sovereignty rests on covenant between vanquished (the prospective subject) and victor (the prospective subject).

(ii) Covenant is mutual transfer of right.

(iii) No right is transferred by the sovereign to the subject.

But the inconsistency disappears, if we replace (ii) by (ii'):

(ii') Covenant is mutual transfer of benefit.

We may suppose that in the usual case transfer of benefit does involve transfer of right, but in the covenant establishing despotic sovereignty the benefit afforded the vanquished—life and corporal liberty—requires no transfer of right on the part of the victor.

Thus although the despotic sovereign is a party to the covenant which establishes his position, this provides no ground for forfeiture of sovereignty because he cannot violate the covenant. And although the subject makes the covenant on conditions, these conditions do not detract from the absolute position of the sovereign, since they are in fact always present in any covenant of obedience—no man obliges himself to obey in the face of death or slavery.

Hobbes's account of despotic sovereignty thus leads to the same conception of sovereign right as his account of instituted sovereignty. The sovereign retains the full right of nature, made effective by the subject's renunciation of right in promising obedience. And one of the major problems of instituted sovereignty is evidently overcome. For we found that Hobbes had no satisfactory way of explaining the subject's obligation *to* the sovereign, and hence no way of denying the possibility that the subjects might depose the sovereign by mutually releasing each other from the covenant instituting him. But in the case of despotic sovereignty the subject's obligation is owed directly to the sovereign, and hence the subjects cannot release themselves from their covenant and so depose their sovereign.

Hobbes's account of despotism fits the picture provided by his initial theory of sovereignty of the relation between sovereign and subject. But in doing so it serves only to bring into sharper relief the inadequacies of that picture. For it shows that the relation of sovereign to subject may truly be compared with that of master and servant. Hobbes, indeed, explicitly draws this comparison (*E.W.* iii, p. 189). But a servant is in no sense positively involved in the household. His will is not included in the will of the master; rather, in relation to the master, the servant has no will. The servant carries out his master's will, differing from a slave only in being allowed corporal liberty. He is not a participant in the household but only its instrument.

If we regard the subject similarly, as the instrument of civil society in the person of the sovereign, then Hobbes's account begins to assume a totalitarian dimension. And this is quite contrary to Hobbes's intention. He succeeds in misleading both himself and his readers by accepting the comparison between master and sovereign, servant and subject. Indeed he invites the rejoinder, urged strongly by Locke,[1] that the sovereign is the enemy of the subjects, and an enemy given the strength to overpower and destroy them by their own act in creating him.

When we have examined the theory of sovereignty in *Leviathan*, we shall be in a position to see why this rejoinder is at least formally misplaced. But it must be admitted that Hobbes never modifies—or can modify—his account of despotic sovereignty to bring it into line with the new picture he develops of the relation between sovereign and subject. He does attempt to use his new political concept, authorization, but to no avail. Despotic dominion is an aberration, in terms of Hobbes's later theory.

Paternal dominion is the other form of sovereignty by nature. Its basis is simple. The child is naturally in the power of its parent, and 'every man is supposed to promise obedience, to him, in whose power it is to save, or destroy him' (*E.W.* iii, p. 188). The child cannot in fact make this promise, but we may suppose it in virtue of our knowledge of human nature, and indeed the parent would have no reason to raise the child, if he were nourishing a prospective enemy rather than a prospective subject.

Hobbes does, however, say elsewhere:

> Over natural fools, children, or madmen, there is no law, no more than over brute beasts; nor are they capable of the title of just, or unjust; because they had never power to make any covenant, or to understand the consequences thereof; and consequently never took upon them to authorize the actions of any sovereign, as they must do that make to themselves a commonwealth. (*E.W.* iii, p. 257)

Perhaps Hobbes's thought is that the child comes to acknowledge its parent as sovereign as it comes to the understanding requisite to make covenants. If it be supposed that this understanding develops while the child is still in the power of the parent, then the child will recognize the necessity of its submission.

[1] Cf. John Locke, *An Essay Concerning . . . Civil Government*, Second Part, section 93.

Within civil society the parent is not sovereign over the child, both being subjects of the civil sovereign. Yet Hobbes supposes that the parent retains some right over the child:

> . . . originally the father of every man was also his sovereign lord, with power over him of life and death; . . . the fathers of families, when by instituting a commonwealth, they resigned that absolute power, yet . . . never intended, they should lose the honour due unto them for their education. For to relinquish such right, was not necessary to the institution of sovereign power; nor would there be any reason, why any man should desire to have children, or to take the care to nourish and instruct them, if they were afterwards to have no other benefit from them, than from other men. (*E.W.* iii, p. 329)

Hobbes's view of family relationships is clearly an unusual one. The primary relation between child and parent rests on fear—the child submits to the parent for fear of the parent's power. Fear, indeed, is the common ground of all human relationships, the basic differentiation being between fear among equals—the relation of men in the state of nature, and fear among unequals. As a species of the latter, the relation of child to parent is comparable with that of servant to master and that of subject to sovereign. It is true that Hobbes supposes sons to be treated more honourably than servants, and ascribes this to the natural indulgence of parents (*E.W.* iv, p. 158). But in terms of rights children and servants are equal.

There is an apparent incompatibility between Hobbes's account of the family and his account of the state of nature. If men are born and raised in families, and if each family is a small-scale society, then how can men ever exist in the condition of war of all against all? In part this question confuses a description of the actual state of affairs (the family) with a logical abstraction used in devising a general theory of human nature (the state of nature). But it must be admitted that the logical device is incompatible with the description, in that it presupposes that men may be treated as fully independent one of another, whereas the existence of the family shows that they are not independent. Hobbes's asocial account of human nature is not an account of the nature of a creature born and raised in a family.

This need not be a damaging criticism of Hobbes's theory. In abstracting from the social characteristics of man, Hobbes sets the extreme case for political theory—the case in which society must be developed for and suited to an asocial being. If this can be done, then it must be possible to show that some kind of society is suited to and so justified for man as he is. If it cannot be done, then it should be possible, by determining why it cannot, to set the minimum degree of sociability required in human nature, if some kind of society is to be suited to and justified for man.

IV

THEORY OF AUTHORIZATION

ONE of the more curious aberrations of Hobbesian scholarship is found in the view that *Leviathan* is 'a rhetorical and, in many ways, a popular *Streitschrift*' in comparison with 'the more calmly argued statements of the same doctrine contained in the *Elements of Law*' (that is, *De Corpore Politico*) 'or the *De Cive*'.[1] There are, it is true, elements of Hobbes's moral doctrine—for example the connection of reason and right—which are stated more clearly in the earlier works. But the political doctrine of *Leviathan* represents a major advance in Hobbes's thinking, an advance which depends almost entirely on the introduction of the concept of authorization.[2]

Yet Hobbes's readers may be forgiven for underestimating the significance of this concept in Hobbes's political thought. It is introduced in the final chapter of the First Part of *Leviathan*, the part entitled *Of Man*. The chapter bears no very obvious relation to what has gone before, nor is there any explanation of the uses to which the concepts it introduces are to be put. It may appear to be somewhat irrelevant appendix or addendum to Hobbes's psychology and moral theory, rather than as the cornerstone of his political theory.

Indeed, those who have taken the chapter seriously have not always construed its significance correctly. Oakeshott, for example, supposes obligation and authorization to be closely linked in Hobbes's argument.[3] We shall later examine his view, and demonstrate its untenability.

---

[1] A. E. Taylor, op. cit., p. 35.

[2] In *De Cive* Hobbes does speak of 'the right and the exercise of supreme authority' (*E.W.* ii, p. 165). But he is concerned there with the exercise of sovereign authority by some other person. This requires authorization *by* the sovereign. He does not introduce authorization *of* the sovereign, i.e. the institution of sovereignty itself by an act of authorization. Authorization by the sovereign is of no particular importance in the structure of Hobbes's theory.

[3] Michael Oakeshott, op. cit., p. lx.

In this chapter we shall try to set out the structure of Hobbes's mature political theory in a manner which will show the essential role played by authorization. We shall employ the method of analysis used in setting out Hobbes's moral theory: first, we shall give formal definitions of Hobbes's primary political concepts; second, we shall give material definitions of these concepts, and indicate briefly the content of the theory; third, we shall discuss a number of problems which arise in stating Hobbes's theory; fourth and finally, we shall attempt to assess the adequacy of the theory, both in terms of Hobbes's aims and in terms of our own interests.

I. THE POLITICAL CONCEPTS: FORMAL DEFINITION

A. *Authorization*

To arrive at a formal definition of authorization, we must proceed through a series of definitions of the concepts found in *Leviathan*, chapter 16, 'Of Persons, Authors, and Things Personated'. The first such concept is that of a person: 'A PERSON, is he, *whose words or actions are considered, either as his own, or as representing the words or actions of another man, or of any other thing, to whom they are attributed, whether truly or by fiction*' (*E.W.* iii, p. 147). Persons, then, fall into two classes: 'When they are considered as his own, then is he called a *natural person*: and when they are considered as representing the words and actions of another, then is he a *feigned* or *artificial person*' (ibid.).

An artificial person is not a non-human entity given in law the status or partial status of a man, but rather a real man or group of men, considered as representing some other man, or group, or thing. If I give you the legal power to conclude an agreement on my behalf, then you in so doing are representing me, and thus you are an artificial person. If you then conclude another agreement on your own behalf, you are a natural person. The distinction between a natural and an artificial person depends solely on whether the man in question is representing himself or another.

The next pair of concepts to be defined are actor and author: 'Of persons artificial, some have their words and actions *owned* by those whom they represent. And then the person is the *actor*; and he that owneth his words and actions, is the AUTHOR: in which case the actor acteth by authority' (*E.W.* iii, p. 148).

Not all things represented or personated are authors. Only that which is itself capable of action—of behaviour resulting from deliberation—can be an author. Many other things can be personated: Hobbes lists 'a church, an hospital, a bridge, . . . children, fools, and madmen that have no use of reason, . . . (a)n idol, or mere figment of the brain' (*E.W.* iii, pp. 149–50). None of these is an author.

It would seem to follow that not all representatives are actors. But Hobbes's account is not clear. That which is not capable of action can be represented only in so far as it is owned, or under the dominion of some man or group. Now if the owner appoints someone else to personate the object, then that appointee is an actor, and the owner is author. But what of the owner himself? May he not represent or personate the thing directly, and not as an actor?

Hobbes insists that those things not capable of action 'cannot be personated, before there be some state of civil government' (ibid.). For in the state of nature there is neither ownership nor dominion. Hence it might be supposed that Hobbes considers the sovereign to be in all cases the author, and the owner merely the actor. And it might be thought that this view is suggested in Hobbes's account of the personation of idols, where he says '(t)he authority proceeded from the state' (ibid.).

But the idols are, presumably, owned directly by the state, or the sovereign. They are public property. Private property is not owned by the state, although the right to private property derives from the sovereign. And it is ownership which determines the author. Hence we conclude that the owner of some object may personate it directly, and not as an actor.

Even if our view is rejected, it will still be important to

distinguish what is involved in being a representative, from what is involved in being an actor. If you appoint me to represent a company which you own, then I am a representative *in relation to* that company, but an actor *in relation to* you, the owner.

A further question, which Hobbes does not consider, is whether a natural person—a man acting on his own behalf—is to be considered an author. There seems to be no reason not to take the signification of author to include this case, and so we shall do so. But nothing in the argument will turn on it.

Since the concepts of actor and author are of great importance in Hobbes's argument, we give their formal definitions in the following equivalences:

'A is an actor' = 'A's words and actions are owned by some other man, or body itself capable of action, B'.

'A is an author' = 'There is some man, or body itself capable of action, B (where B may = A), whose words and actions are owned by A'.

The next concept we require is authority: 'And as the right of possession, is called dominion; so the right of doing any action, is called AUTHORITY. So that by authority, is always understood a right of doing any act; and *done by authority*, done by commission, or licence from him whose right it is' (*E.W.* iii, p. 148).

From this passage we may derive two equivalences:

'A has the authority to do X' = 'A has the right to do X'.

'A does X by authority of B' = 'B has the right to do X, and B has granted the right to do X to A, and A does X'.

Presumably a man may act by his own authority, in which case the second clause in the definition, which becomes 'A has assigned the right to do X to A', is unnecessary. Note also that 'right' in these equivalences is to be understood, in its usual Hobbesian sense, as merely permissive. There is no problem in handling either exclusive rights, or rights entailing obligations on others, simply by appropriate expansions of both sides of the equivalences.

We come now to authorization. Hobbes does not provide an explicit definition, and indeed the word 'authorization' and the related verb 'authorize' provide only one of the ways in which he expresses the concept. To authorize someone is to appoint him one's representative, or to appoint him to bear one's person, or to give him the right to present one's person, or to give him one's authority. Authorization seems the most convenient label for this procedure, and our problem is to determine precisely what it involves.

From the last equivalence above it is clear that authorization must involve some translation of right. This is evidently not mere renunciation, nor is it transfer, in Hobbes's usual sense. For transfer is nothing more than a limited renunciation; if I transfer my right to some object to you, I merely renounce my right to possess that object in so far as this would interfere with the exercise of your right to the object. Authorization, on the other hand, enables *you* to act in my place, and so with *my* right.

Thus although the right of nature is unlimited, authorization does confer upon the actor a right which he did not previously possess. For the right of nature is unlimited in scope, but not in person. One man's right is distinct from another man's; hence the right of nature does not enable one man to act for another—to perform actions owned by the other. Authorization, however, does enable one man to act for another. One man's right *becomes* another man's right, in the sense that the actor gains *the use of* the author's right.

We propose, then, the following definition of authorization:

'authorization' = 'that procedure by which some man (or body capable of action), A, gives the use of (some portion of) his right, to some other man (or body capable of action), B'.

Leaving out the various qualifications, we may say for short:

'authorization' = 'that procedure by which one man gives the use of his right to another man'.

Two questions immediately suggest themselves. Does the procedure of authorization require that the use given be either exclusive or permanent? That is, should we say in the definition, not just 'gives the use of his right' but 'gives the exclusive and/or permanent use of his right'?

There is surely no reason to take authorization to be permanent. If it is intended to cover that commonplace procedure whereby I make someone else my agent, then it must be possible for authorization to be withdrawn as well as granted. We shall, therefore, assume that authorization can be revoked or withdrawn. This will have important consequences for our discussion of the relation between authorization and obligation.

But is authorization, while it lasts, exclusive? More particularly, if A gives B the use of his right, must A forgo the use of his right, as long as his authorization of B stands? The answer seems to be in principle negative. Of course, if the effect of the authorization would be undone by A's exercise of his right, then of course he must forgo it as long as his authorization stands—but this is a purely logical consideration. In other cases there seems to be no reason why A cannot exercise the right he has also authorized B to exercise. Indeed, Hobbes says later: 'Again, the consent of a subject to sovereign power, is contained in these words, *I authorize, or take upon me, all his actions*; in which there is no restriction at all, of his own former natural liberty: . . . ' (*E.W.* iii, p. 204). I take it that Hobbes intends us to understand that by authorizing another man, we do not in any sense forgo our own rights.

Finally, we may consider briefly the connection between authorization and responsibility. Since whatever the actor does, within the scope of his authorization, is to be regarded as the act of the author, then the actor is responsible for them. If then the actor makes a covenant, acting within the scope of his authorization, the author is put under obligation by it.

Should the actor exceed the scope of his authorization, then he, and not the author, bears the responsibility. Should he act on improper authorization—that is, on authorization which the author lacked right to give—Hobbes's doctrine seems to be that the author alone is responsible if the actor is obliged to obey the author; otherwise, the two must presumably share responsibility (cf. *E.W.* iii, p. 149).

Hobbes's concept of authorization seems straightforward and, one might say, quite innocuous. It has no obvious relation to Hobbes's moral concepts, and it is an unpromising candidate for the cornerstone of an absolutist political theory. What Hobbes does with it seems—and is—little short of amazing.

B. *Sovereign and Sovereign Right*

Hobbes says that

the essence of the commonwealth . . . is *one person, of whose acts a great multitude, by mutual covenants one with another, have made themselves every one the author, to the end he may use the strength and means of them all, as he shall think expedient, for their peace and common defence.*

And he that carrieth this person is called SOVEREIGN, . . . (*E.W.* iii, p. 158)

To extract a formal definition of the sovereign from this passage we must omit all reference to the end—peace and common defence—since this belongs properly to the material content of Hobbes's political theory. We must note also that the sovereign must be the only actor given the use of the rights he possesses, and that he may be either one man, or an assembly of man. An assembly may be defined as a number of men, with a procedure for common deliberation and action.

We now propose this definition:

'sovereign' = 'an actor, either a man or an assembly, who has the use of the several rights of a number of men, given only to him by covenant among themselves'.

The success of Hobbes's new conceptual departure in meeting one of the problems in the last chapter, and in presenting a new picture of civil society, is at once apparent. The sovereign is no longer conceived as a natural person, possessed of the full right of nature made effective by the renunciation of natural right on the part of others, but an artificial person, possessed of the use of each man's right of nature. Every man is thus evidently involved in society in a positive manner, for the acts of the sovereign may be considered his own acts. The sovereign, in Hobbes's mature theory, is conceived as the representative of every member of society.

We postpone for the present the problems to which this new account gives rise. Proceeding with the basic concepts, we must attempt to define sovereign right (which Hobbes, with his usual lack of care in this matter, does not distinguish from sovereign power).

In his account of sovereignty by institution, Hobbes speaks of the sovereign being given 'the *right* to *present* the person of them all' (*E.W.* iii, p. 159). That is, the *use* of each man's right may be itself considered a right, a right to present the person of each man, or as we might say, a right to represent them, or again, a right to perform actions which each man owns. This may suggest the first stage of a definition of sovereign right:

'sovereign right' = 'the right of an actor to perform acts owned by each of a body of men'.

But there are several inadequacies in this equivalence. In the first place, it fails to show that only one man or assembly can possess sovereign right with respect to any body of men. Second, it fails to show that sovereign right is effective— that it may not be opposed by those who have authorized the sovereign. And third, it fails to show that sovereignty is permanent—that it cannot be terminated at the will of the subjects.

The first inadequacy may easily be remedied. The third may be remedied at least in part, by introducing into the definition reference to the fact that the subjects are obliged by covenant among themselves to maintain their authorization of the sovereign. The possibility that the subjects may mutually dissolve their covenant—a possibility which as we have seen Hobbes denies—cannot be eliminated here by any simple alteration of the definition. But it will be discussed later, when we shall consider if Hobbes can eliminate it in any manner consistent with his general theory.

The second inadequacy may seem most formidable. For we insisted in the preceding section that authorization does not deprive the author of the right whose use is conferred upon the agent. Surely this means that those who authorize the sovereign retain their full right of nature, and so may oppose him and prevent his right from being effective.

But we also pointed out, as a purely logical consideration, that as long as the authorization stands, the author must forgo any right which would undo the effect of the authorization. He *can*, that is, do nothing incompatible with his authorization. And this is here made effective by the obligation to maintain the authorization which each of the authors assumes by covenant. That is, each can do nothing incompatible with the authorization as long as it stands, and each has an obligation to allow it to stand. Therefore each *may* do nothing incompatible with his authorization. Whether this suffices to give the sovereign effective right is another question we postpone for later consideration. But our definition now reads:

'sovereign right' = 'the right, belonging exclusively to one actor, to perform acts owned by each of a body of men, effective, in so far as each may do nothing incompatible with his authorization so long as it stands, and permanent, in so far as each is obliged by covenant among themselves to maintain his authorization'.

The sovereign, then, is an actor, but an actor whom the authors have an obligation one to another to maintain, and hence an actor whom the authors have an obligation not to oppose in any manner incompatible with the maintenance of their authorization. What this implies, in terms of determinate obligations, depends of course on the content of the act of authorization, which will concern us in the next part of this chapter. But it is clear that, although a representative, the sovereign is a permanent representative with certain rights over those whom he represents.

C. *Subject and Liberty of Subjects*

The definitions in this section should be the correlates of the definitions in the preceding section. As we shall see, however, Hobbes's account does not always make the correlation evident.

The prospective subjects must

> confer all their power and strength upon one man, or upon one assembly of men, that may reduce all their wills, by plurality of voices, unto one will: which is as much as to say, to appoint one man, or assembly of men, to bear their person; and every one to own, and acknowledge himself to be author of whatsoever he that so beareth their person, shall act, or cause to be acted, in those things which concern the common peace and safety; and therein to submit their wills, every one to his will, and their judgments, to his judgment. This is more than consent, or concord; it is a real unity of them all, in one and the same person, made by covenant of every man with every man, in such manner, as if every man should say to every man, *I authorise and give up my right of governing myself, to this man, or to this assembly of men, on this condition, that thou give up thy right to him, and authorize all his actions in like manner.* (*E.W.* iii, pp. 157–8)

Certain features of this account do not quite correspond to the argument we have been setting out—in particular the use of the phrase 'I authorise and give up my right'. If our account of authorization is correct, it does not involve the giving *up* of right. However, we should note that in so far as the exercise of some right is incompatible with the authorization, then that right must in effect be given up, and we may perhaps suppose that this is what Hobbes intends here.

Thus we propose this definition:

'subject' = 'an individual author, who is one of a number of men who have agreed, by mutual covenant, to give the use of their several rights to some one man, or assembly'.

Hobbes's account of the liberty of subjects has two parts. First, there are 'the things, which though commanded by the sovereign, he may nevertheless, without injustice, refuse to do' (*E.W.* iii, p. 203). Second, there are those liberties which 'depend on the silence of the law' (*E.W.* iii, p. 206). What these two types of liberty are we shall consider in our examination of the material definitions of Hobbes's concepts.

Since 'CIVIL LAW, *is to every subject, those rules, which the commonwealth hath commanded him*' (*E.W.* iii, p. 251), we may express our definition of the liberty of subjects by this equivalence:

(I) 'A subject, A, has liberty to do X' = 'Either A has the right to do X even though doing X may be forbidden by the sovereign, or doing X is not forbidden by the sovereign'.

But it seems possible to define the liberty of subjects in a rather different way.[1] In line with our previous investigation into liberty, the liberty of subjects would seem to be coextensive with, or even equivalent to, the rights of subjects. Now these rights extend, presumably, to whatever rights are not given up in the covenant to authorize the sovereign. Authorization itself requires only that one forgo any right incompatible with the authorization, and since the authorization is obligatory, such rights are not part of the rights of subjects. All other rights, then, are included in the rights of subjects.

One qualification is needed, even to the supposition that the only rights forgone are those incompatible with the authorization. It may be that under certain conditions a subject may renounce his authorization, despite his covenant to maintain it. What these are is a material question, but of course we know that for Hobbes the subject always retains the right to do what is directly necessary for his survival. However, this material condition does not itself enter the present definition, which is given by the equivalence:

(II) 'A subject, A, has liberty to do X' = 'Either A has the

---

[1] This second way is suggested by Hobbes himself, who proposes to determine 'the things, which though commanded by the sovereign, he [the subject] may nevertheless, without injustice, refuse to do' by considering 'what rights we pass away, when we make a commonwealth; or, which is all one, what liberty we deny ourselves, by owning all the actions, without exception, of the man, or assembly we make our sovereign' (*E.W.* iii, p. 203).

right to do X even though doing X is incompatible with
A's authorization of the sovereign, or doing X is not incompatible with A's authorization of the sovereign'.

We may use (II) to show the inadequacy of (I). Let Y
be some action incompatible with A's authorization of the
sovereign, but not such that A has the right to do it notwithstanding its incompatibility. Suppose, however, that the
sovereign has failed to forbid A to do Y. Then according to
(I) A is at liberty to do Y, although according to (II) he is not.
In such a case we surely want to uphold (II); we should not
say that the sovereign's failure to forbid some act incompatible with his authorization gives his subjects liberty to perform it.

Whether (I) can be used to show the inadequacy of (II)
depends on the interpretation of incompatibility. Let Z be
some action forbidden by the sovereign, but not such that A
has the right to do it notwithstanding its being forbidden.
Then unless we suppose that *therefore* doing Z is incompatible
with A's authorization of the sovereign, it may be compatible.
If it is, then according to (II) A is at liberty to do Z, although
according to (I) he is not. In such a case we want to uphold (I);
if the subjects do not have the right to ignore the sovereign's
command, then even though the act forbidden may not be
incompatible with their authorization, they cannot have the
liberty to perform it.

Leaving open the issue raised in the last paragraph, we
propose the following combined definition:

'A subject, A, has liberty to do X' = 'Either A has the right
to do X even though doing X may be forbidden by the
sovereign, or doing X is not incompatible with A's
authorization of the sovereign and doing X is not forbidden by the sovereign'.[1]

[1] The definition is derived as follows. Let 'A has the right to do X even though
doing X may be forbidden by the sovereign' be 'p'. Let 'doing X is not forbidden
by the sovereign' be 'q'. Let 'A has the right to do X even though doing X is
incompatible with A's authorization of the sovereign' be 'r'. Let 'doing X is not

The last clause may be omitted if we suppose that any act not satisfying the first condition is such that if it is forbidden by the sovereign, it is incompatible with authorization of the sovereign.

D. *Conclusions*

Just as the formal definitions of Hobbes's moral concepts do not depend on his account of human nature, so the formal definitions of his political concepts are psychologically neutral. They provide an abstract schema, into which we can put content by making certain suppositions about the purpose of the act whereby the sovereign is authorized. And for Hobbes these suppositions depend on his view of man.

The formal definitions thus do not dictate the content of the theory. Nothing in them prevents authorization of the sovereign being limited, as authorization normally is. We do not usually authorize some person or group to use *all* of our rights. And a limited authorization would establish only a limited sovereign right, and consequently the covenant to authorize would impose only a limited obligation on the subjects.

We emphasize this because much of the interest in Hobbes's moral and political theories is to be found not in their content, but rather in their conceptual structure. That this has been frequently overlooked is understandable, because it is the content that makes the most striking impression upon the reader, and this is certainly what Hobbes intended. Yet it is reasonable for us to be ultimately more impressed by other

incompatible with A's authorization of the sovereign' be 's'. Let 'a subject, A, has liberty to do X' be 't'.

Then it seems clear that we may assume:

   (i) if p, then t
  (ii) if q and s, then t
 (iii) if t, then p or q
 (iv) if t, then r or s
  (v) if r, then p

From these we may derive:

 (vi) t, if and only if, p or (s and q)

And this is our definition.

features of Hobbes's argument. The content is, as we have shown in the case of the moral theory and will show in the political theory, inadequate. But the conceptual structure may provide us with insights into the construction of a more adequate theory.

Such a theory will of course not be developed in this book. At best we can make a few suggestions towards it. But the importance of examining Hobbes's arguments is surely not merely historical; the labour which has recently been expended on his works would hardly be justified by a gain only in historical insight. The underlying rationale of our study is that Hobbes has something of value for us, confronted with moral and political problems, even if what is of most value is not what he intended.

The conclusions we draw from our formal inquiry are these:

1. The concept of authorization provides an illuminating and fruitful way of expressing the relation between sovereign (or government) and subjects (or citizens).
2. The extent of the authorization granted to the sovereign determines the extent of sovereign right.
3. The obligation on the subjects to maintain this authorization determines the extent of their liberty.

In the next two parts of this chapter we shall determine the implications, for Hobbes, of the last two of these conclusions. In the final part we shall develop the first conclusion, and assess Hobbes's success in providing a political solution to the problems of the state of nature.

2. THE POLITICAL CONCEPTS: MATERIAL DEFINITION

A. *Authorization*

Hobbes's treatment of authorization is strictly formal. We could expand the formal definition given in the last part by adding a clause stating that authorization must be in accord with right reason. And we could then transform this into a material definition by replacing this clause with one stating the actual requirements of right reason. But these definitions would add nothing of value to our analysis, and would only serve to complicate it unnecessarily. We therefore shall not pursue further the material definition of authorization.

B. *Sovereign and Sovereign Right*

The material definition of the sovereign may be obtained from the formal by adding a clause specifying the use to be made, by the sovereign, of the authorization given him. Hobbes speaks of the 'peace and defence' of the subjects, or 'the common peace and safety' (*E.W.* iii, p. 158). It will be useful, however, to consider the role of the sovereign as two-fold—to make the subjects secure against one another, and to make them secure against external enemies. This corresponds to the basic division between the sword of justice and the sword of war, developed in the earlier political writings. It distinguishes the internal role of the sovereign from the external role. Thus we propose the following definition:

> 'sovereign' M = 'an actor, either a man or an assembly, who has the use of the several rights of a number of men, for the purpose of making them secure against each other, and against external enemies, given only to him by covenant among themselves'.

And the definition of sovereign right follows immediately by a similar addition:

> 'sovereign right' M = 'the right, belonging exclusively to one actor, to perform acts owned by each of a body of men, for the purpose of making them secure against each other, and against external enemies; effective, in so far as each may do nothing incompatible with his authorization so long as it stands, and permanent, in so far as each is obliged by covenant among themselves to maintain his authorization'.

Hobbes does not develop his account of the extent of sovereign right as systematically in *Leviathan* as in the earlier works. But his doctrine is nevertheless clear; sovereign right is unlimited. For he says 'the consent of a subject to sovereign power, is contained in these words, *I authorize, or take upon me, all his actions*' (*E.W.* iii, p. 204). Now it is not clear how this extreme position can be maintained; the sovereign has the right, in Hobbes's view, to kill the subject, but how can the subject authorize the sovereign—that is, give him the use of his own right—to do this? But this is a problem to which we shall return; it suffices for present purposes to note that in Hobbes's view the authorization given by the subject, and so the right acquired by the sovereign, is as extensive as possible.

But can Hobbes justify this? Need the sovereign be handed a blank cheque by his subjects? Hobbes's answer is found in this passage:

> The opinion that any monarch receiveth his power by covenant, that is to say, on condition, proceedeth from want of understanding this easy truth, that covenants being but words and breath, have no force to oblige, contain, constrain, or protect any man, but what it has from the public sword; that is, from the untied hands of that man, or assembly of men that hath the sovereignty, and whose actions are avouched by them all, and performed by the strength of them all, in him united. (*E.W.* iii, p. 162)

The condition Hobbes considers in this passage—the condition that would be imposed by making the sovereign a real party to the covenant, transferring certain rights to his subjects—is not the one we are examining here. We are concerned with a limited act of authorization. But the effect would be the same, in that it would limit sovereign right. And this, Hobbes insists, is to deprive the sovereign of the means necessary to uphold covenants, and so guarantee the internal security of society.

In chapter 29, 'Of those things that weaken, or tend to the Dissolution of a Commonwealth', Hobbes places first '*Want of absolute power*' (*E.W.* iii, p. 309). And since the maintenance of commonwealth is man's only security against a return to the state of nature, absolute power must be considered, not only necessary, but good, as one of the means of peace. Note, of course, that it is *power* and not *right* which concerns Hobbes here. Whether the power of the sovereign can in fact be absolute is a question we shall face in our final assessment of Hobbes's political theory. But if the sovereign requires absolute power, then he must be supposed to have unlimited right, since the extent of his right is determined by what is necessary for him to do what he is authorized to do.

Hobbes thus concludes:

> So that it appeareth plainly, to my understanding, both from reason, and Scripture, that the sovereign power . . . is as great, as possibly men can be imagined to make it. And though of so unlimited a power, men may fancy many evil consequences, yet the consequences of the want of it, which is perpetual war of every man against his neighbour, are much worse. (*E.W.* iii, pp. 194–5)

In an even stronger passage, Hobbes objects to those who criticize the unlimited extent of sovereign power and right, that they do not consider

> that the state of man can never be without some incommodity or other; and that the greatest, that in any form of government can possibly happen to the people in general, is scarce sensible, in respect of the miseries, and horrible calamities, that accompany a civil war, or that dissolute condition of masterless men, without subjection to laws, and a coercive power to tie their hands from rapine and revenge (*E.W.* iii, p. 170)

Whether the totalitarian governments of our acquaintance would have led Hobbes to change his doctrine, we can only speculate.

Hobbes has a further argument in support of the unlimited right of the sovereign. Not only is it desirable, but it is also, Hobbes thinks, unavoidable. For once sovereign power is established, then 'whosoever thinking sovereign power too great, will seek to make it less, must subject himself, to the power, that can limit it; that is to say, to a greater' (*E.W.* iii, p. 195).

This argument seems to depend on two suppositions. The first is of course basic to Hobbes's whole account of sovereignty —that the sovereign, by virtue of his right, can obtain some power superior to that of the individual in the state of nature. And this surely holds, even if sovereign right is not acknowledged to be unlimited. The second is that if one power is clearly superior to another, it is effectively absolute with respect to that other power. This is not stated, much less argued, but it is hard to see how any lesser supposition could suffice for the argument Hobbes presents. If a lesser power cannot of itself limit a greater, then that greater must be effectively absolute in relation to the lesser.

If we grant Hobbes this supposition, then not only are all problems about making the sovereign's right effective solved, but also all arguments about the desirability of unlimited sovereign right are made redundant. For if a greater power is always effectively absolute, then it can always compel recognition of its right, by threatening death as the alternative. A sovereign by institution, whatever his rights may be by the act of institution, will always be able to convert himself into a sovereign by acquisition, with unlimited right.

The difficulty is, of course, that there seems not the slightest reason to grant Hobbes this supposition, either as part of his theory or as an independent, well-grounded claim. Hobbes's problem is to ensure sufficient power to the sovereign—as he recognizes in his criticisms of those who would limit sovereign power. On this supposition, however, they would be attempting not just what Hobbes considers undesirable, but what is impossible. We therefore cannot consider this further argument to be consistent with the general character of Hobbes's political theory, although we shall want to return to it in our final assessment of the theory.

Since the act of authorization confers upon the sovereign the fullest possible use of the rights of his subjects, it follows that the sovereign cannot forfeit his position, that he can do no injustice to his subjects (for they own his acts, and no man can be unjust to himself), that he determines the means of peace and defence and makes war or peace, that he judges what instruction shall be given his subjects and what views may be expressed by them, that he determines property-rights and property disputes, that he makes laws, chooses his own ministers, rewards and punishes.[1] Moreover, all these rights of the sovereign are indivisible; for 'powers divided mutually destroy each other' (*E.W.* iii, p. 313).

We shall consider in the next part of this chapter the right to punish, since it is not easy to see how this may be derived from authorization—no man having the right to inflict harm on himself. But the other sovereign rights seem to follow well enough, once it is granted that the purpose of instituting sovereignty requires that sovereign right be unlimited.

The unlimited extent of sovereign right has occasioned much unfavourable reaction from Hobbes's critics. It must be remembered, of course, that this right belongs to an *artificial* person, but since an artificial person is an individual or assembly in its capacity as representative, the critics may justly insist that sovereign right is conferred upon an actual man or men, with all the passions and appetites that Hobbes finds in human nature.

For this there is no remedy. Hobbes can make the sovereign representative of his subjects, but he cannot avoid making him also a natural body—a man, or an assembly of men. Yet he is concerned, more in *Leviathan* than in his earlier works, with those *institutional* devices within which men may live at peace one with another, and which, although he does not dwell on this, in fact must modify the exercise of sovereign power, even if they leave untouched sovereign right.

In *De Corpore Politico* and *De Cive* Hobbes emphasizes the sovereign's role in wielding the swords of justice and of war. In *Leviathan* he tends rather to emphasize the sovereign's role in establishing and maintaining those social institutions to which we have just referred. Neither of the earlier writings contains most of the subject-matter to be found in chapters 22 to 25 and 28 of *Leviathan*—'Of Systems Subject, Political and

---

[1] Detailed references to the sources of this paragraph seem unnecessary. The various rights ascribed to the sovereign are found, similarly set out, in chapter 18 of *Leviathan*.

Private'—which treats of municipal corporations, trading associations, and other bodies within the state; 'Of the Public Ministers of Sovereign Power'; 'Of the Nutrition, and Procreation of a Commonwealth'; 'Of Counsel'; 'Of Punishments and Rewards'.

These chapters are commonly neglected by Hobbes's critics. No doubt a detailed analysis of their arguments is unnecessary to an understanding of the structure of Hobbes's political theory, but their general character casts an important light on what Hobbes deemed to be the nature of the society to which his theory applied. It becomes evident, on reading these chapters, that Hobbes intends no totalitarian system, or arbitrary despotism, but rather an enlightened monarchy, authoritarian but benevolent, offering the subjects both ample opportunity to make known their needs and grievances before the seats of power, and adequate freedom to engage in commercial and mercantile activities.

Hobbes's critics never tire of pointing out that, according to his theory, no law can be unjust. But in one of the very passages in which Hobbes maintains this, he also is concerned to define a good law, as 'that, which is *needful*, for the *good of the people*, and withal *perspicuous*' (*E.W.* iii, p. 335).

And the best counsel upon which to base these laws 'is to be taken from the general informations, and complaints of the people of each province, who are best acquainted with their own wants, and ought therefore, when they demand nothing in derogation of the essential rights of sovereignty, to be diligently taken notice of' (*E.W.* iii, p. 341).

In *Leviathan* the positive role of the sovereign, implicit in the view of the sovereign as actor authorized by his subjects, is given content. To ignore this content because it is not part of the logic of sovereignty—because a sovereign who neglects his duties[1] does not thereby violate any obligation to his subjects—is to deny Hobbes a fair hearing.

---

[1] Warrender finds it an 'exacting proposition that the sovereign may have duties, but has no obligations' (op. cit., pp. 156–7). But Oakeshott, who is here being criticized, is entirely correct in his view of the sovereign's duties. They

C. *Subject and Liberty of Subjects*

The material definition of subject is obtained from the formal definition by the addition of the same clause which distinguishes the formal and material definitions of sovereign:

> 'subject' M = 'an individual author, who is one of a number of men who have agreed, by mutual covenant, to give the use of their several rights to some one man, or assembly, for the purpose of making them secure against each other, and against external enemies'.

It does not seem possible to give a material definition of the liberty of subjects by a correspondingly simple addition to the formal definition. We shall therefore attempt to arrive at our definition by examining Hobbes's account of the liberty of subjects.

The initial premiss of this account is that the liberties and obligations of the subject are correlative. Hence we consider 'what rights we pass away, when we make a commonwealth' (*E.W.* iii, p. 203); those rights which remain constitute the liberty of subjects. Arguments to determine what rights are given up 'must either be drawn from the express words, *I authorize all his actions*, or from the intention of him that submitteth himself to his power, which intention is to be understood by the end for which he so submitteth'. Thus 'the obligation, and liberty of the subject, is to be derived, either from those words, or others equivalent; or else from the end of the institution of sovereignty, namely the peace of the subjects within themselves, and their defence against a common enemy' (ibid.).

The distinction Hobbes intends here is between what (if anything) can be derived from the act of authorization itself,

---

are 'what it *must* do. Its duties are derived from the end for which it was instituted' (op. cit., p. xl). Hobbes uses 'duty' in this sense (*E.W.* iii, p. 323).

Of course the sovereign does have obligations to God, although not to his subjects, in Hobbes's view. But the existence of these obligations is in no way required for him to have duties, in the sense just given. *Any* office has duties. The holder of the office may have an obligation to perform these duties, but that is a separate matter.

and what can be derived from the covenant obliging us to maintain the authorization. This obligation is, as we have seen, to do nothing incompatible with the authorization, and what is incompatible depends on the *intention* of the authorization. Thus Hobbes speaks on the one hand of the 'express words' (the act of authorization itself) and on the other hand of 'the end of the institution of sovereignty' (which is the subjects' mutual intention). By 'intention', it should be noted, Hobbes does not mean something mental or private, but rather the public aim of the act of authorization, which we have stated in the material definitions of sovereign and subject.[1]

Hobbes then correctly points out that nothing whatever can be derived from the words themselves—the act of authorization is not obliging. We have quoted the passage in which he says this, but we may usefully include it here: 'Again, the consent of a subject to sovereign power, is contained in these words, *I authorize, or take upon me, all his actions*; in which there is no restriction at all, of his own former natural liberty: . . .' (*E.W.* iii, p. 204).

It is, therefore, to the intention of the act of authorization that we must turn, in order to determine the content of the subject's obligations and liberties. Here two principal considerations emerge (*E.W.* iii, pp. 204–5):

1. The subjects cannot be obliged to kill or injure themselves, or not to defend themselves, under any circumstances. This is guaranteed by the inalienable core of the right of nature.

2. The subjects cannot be obliged to undertake any dangerous task, such as killing another, or engaging in warfare, unless the survival of the commonwealth depends on it. That is, the inalienable right of nature must include the right to choose the course of action least dangerous to survival; thus the subjects may be obliged to face danger (war) if a greater danger (destruction of society) would follow if they did not, but not otherwise.

[1] There are other interpretations of this very important argument. In the next part of this chapter we shall discuss the way in which authorization, covenant, and obligation are related, and there we shall treat Warrender's interpretation.

To these considerations Hobbes later adds a third (*E.W.* iii, p. 206):

3. The subjects are not obliged to avoid what the sovereign has not forbidden by law. No doubt we may understand this with the *caveat* that the liberty so allowed does not extend to acts destructive of the commonwealth, such as rebellion, even if these have not been forbidden.

These three classes may be fitted nicely into the formal definition of liberty. The first two classes include what the subjects cannot be obliged to do even on command of the sovereign. Thus these provide the liberties which fall under the first part of the definition. And we might note further that these liberties could be broken down: the first group will include those acts which the subject has the right to do even though they may be incompatible with his authorization of the sovereign; the second group includes those acts which the subject has the right to do as long as they are compatible with that authorization.

The third class includes what the subjects are not obliged to do because the acts are required neither to maintain the authorization of the sovereign nor to conform to the commands of the sovereign. Thus these provide the liberties which fall under the second part of the definition.

We propose, then, the following material definition:

'A subject, A, has liberty to do X' M = 'Either A has the right to do X in order to preserve himself, or A has the right to do X in order to avoid the greatest present danger to his preservation, or A has the right to do X because it is neither incompatible with his authorization of the sovereign nor forbidden by (civil) law'.

The first two parts correspond to the first part of the formal definition, and the third part to the second part of the formal definition.

It is clear from this definition that the most considerable part of the liberty of the subject depends on the silence of the law. Since the extent of the law is determined by the sovereign, liberty may vary from one commonwealth to another, and from one time to another (*E.W.* iii, p. 206). The subject, of course, has no right to complain if his liberty seems unduly narrow.

Furthermore, the '(l)iberty of the subject [is] consistent with the unlimited power of the sovereign' (*E.W.* iii, p. 200). Thus a subject who has not violated the law may be put to death, or imprisoned, or banished, by the sovereign, without injustice, although the subject did only what he had liberty to do: '. . . the people of Athens, . . . sometimes banished an Aristides, for his reputation of justice; and sometimes a scurrilous jester, as Hyperbolus, to make a jest of it. And yet a man cannot say, the sovereign people of Athens wanted right to banish them; or an Athenian the liberty to jest, or to be just' (*E.W.* iii, pp. 200–1).

If the sovereign may ignore his laws with impunity, then what is the significance of liberty? Hobbes's critics have fastened on passages such as that we have just quoted, and attacked Hobbes as the enemy of liberty. And such attacks are not altogether unwarranted—for Hobbes has and can have no place in his theory for denying the overriding right of the sovereign.

Yet we must read these passages against those which emphasize the duty of the sovereign to respect liberty. 'The commodity of living consisteth in liberty and wealth', so that it is 'the duty of a sovereign by the law of nature' to impose 'no restraint of natural liberty, but what is necessary for the good of the commonwealth' (*E.W.* iv, p. 215). In all of his political writings Hobbes insists that liberty is a good, and that the sovereign should not deprive men of it except for a greater good.

Far from being the enemy of liberty, Hobbes would consider himself its true friend. For he recognizes that in the state of nature, where liberty is unlimited, it is also of no value—for no man can exercise his liberty to make himself secure and to live well. 'Every man indeed out of the state of civil government hath a most entire, but unfruitful liberty;

. . .' (*E.W.* ii, pp. 126–7). It is only by limiting 'the natural liberty of particular men, in such manner, as they might not hurt, but assist one another, and join together against a common enemy' (*E.W.* iii, p. 254), that the liberty which remains is of value, because fruitful.

There is a passage in *De Cive* which, although it does not contain the word 'liberty', brings out Hobbes's doctrine most clearly.

> Out of this state [of civil society, ruled by law], every man hath such a right to all, as yet he can enjoy nothing; in it, each one securely enjoys his limited right. Out of it, any man may rightly spoil or kill another; in it, none but one. Out of it, we are protected by our own forces; in it, by the power of all. Out of it, no man is sure of the fruit of his labours; in it, all men are. (*E.W.* ii, p. 127)

It is true that in Hobbes's political system, security comes before liberty. Hobbes would have insisted that liberty depends on security, of course. It is true that his political theory centres on security, rather than liberty, and that in trying to devise guarantees of security, Hobbes must exclude guarantees of liberty. It is true that Hobbes feared anarchy more than tyranny. Yet for all this Hobbes advocates liberty, and he realizes, what many of his contemporaries and successors did not, that liberty is primarily desirable as an instrument—that men want to be free to achieve their ends.

Thus Hobbes realizes that the state must provide those conditions under which men can achieve their ends. His state is more than a policeman. He advocates social welfare and public measures to ensure full employment (*E.W.* iii, pp. 334–5). He has, of course, none of the economic understanding which would be needed to qualify him as one of the fathers of the welfare state. But his political doctrine has greater affinities with the liberalism of the twentieth century than his authoritarian theory would initially suggest.

D. *Conclusions*

In this part of our study we have shown that Hobbes is able to use his account of the end or purpose of commonwealth, and of the means necessary to that end—an account which depends in its turn on his view of human nature—to show how by an act of authorization, which the subjects are obliged by covenant to maintain, a representative is appointed who is in fact supreme over those whom he represents.

Different premises would, of course, have led to a different conclusion. In particular, if the way to peace is not as difficult as Hobbes supposes, then a limited act of authorization may serve to erect a sovereign with sufficient right and power to achieve the purpose of the commonwealth. Such a sovereign would be a representative with only certain limited rights over those whom he represents.

It is impossible to emphasize too strongly that it is the substantive premises about human nature, and not the formal structure of the theory, that determines its absolutist character. If our attempt to distinguish formal from material considerations provides nothing but a recognition of this one fact, it will have been sufficiently justified.

There are one or two minor points about the content of Hobbes's later political theory which might usefully be mentioned at this point. The assumption of the temporal and logical priority of democracy, which we found in the initial theory, is abandoned. The act of authorization may single out an individual from the outset, just as easily as it may authorize an assembly.

Nothing in the theory turns on whether the sovereign is one man or a group, or whether this group is an aristocratic or a democratic assembly. Hobbes has evident preferences for monarchy, but these play no part in the theory, nor have they been neglected in other treatments of Hobbes's political theory. Hence they need no emphasis here.

There is, however, one small but interesting point, about the relation of liberty to the several forms of sovereignty, which is worth noticing. In *De Corpore Politico* Hobbes maintains:

Now seeing freedom cannot stand together with subjection, liberty in a commonwealth is nothing but government and rule, which because it cannot be divided, men must expect in common; and that can be no where but in the popular state, or democracy. And Aristotle saith well, (lib. VI, cap. 2 of his *Politics*), *The ground or intention of a democracy, is liberty.* (*E.W.* iv, p. 202)

In *Leviathan* we find, however, the following:

And because the Athenians were taught, to keep them from desire of changing their government, that they were freemen, and all that lived under monarchy were slaves; therefore Aristotle puts it down in his *Politics*, (*lib.* 6. *cap.* ii.) *In democracy,* LIBERTY *is to be supposed: for it is commonly held, that no man is* FREE *in any other government.* And as Aristotle; so Cicero, ... And by reading of these Greek, and Latin authors, men from their childhood have gotten a habit, under a false show of liberty, of favouring tumults, and of licentious controlling the actions of their sovereigns, and again of controlling those controllers; with the effusion of so much blood, as I think I may truly say, there was never any thing so dearly bought, as these western parts have bought the learning of the Greek and Latin tonuges. (*E.W.* iii, pp. 202–3)

De Cive is more temperate, but supports the position of *Leviathan*, accusing Aristotle of 'miscalling dominion liberty', and terming his quotation '*a speech of the vulgar*' (*E.W.* ii, p. 135). Hobbes rarely shows such a marked change of opinion in his political writings.

3. THE POLITICAL CONCEPTS: SOME PROBLEMS

A. *Authorization and the Sovereign Right to Punish*

In basing his later political theory on authorization, Hobbes provides himself with a serious problem in explaining the basis of the sovereign right to inflict punishment on the subjects. Hobbes begins by failing to notice the problem—by supposing that, in owning all of the actions of the sovereign, the subjects therefore own the acts whereby they are punished. But this, Hobbes later recognizes, is unsatisfactory. He attempts an explanation of the right to punish which, we shall show, is unsuccessful. But we shall be able to construct a better explanation, which will serve Hobbes's purposes.

In arguing that the subjects do not have the right to change the form of government, Hobbes reasons in part as follows:

Besides, if he that attempteth to depose his sovereign, be killed, or punished by him for such attempt, he is author of his own punishment, as being by the institution, author of all his sovereign shall do: and because it is injustice for a man to do anything, for which he may be punished by his own authority, he is also upon that title, unjust. (*E.W.* iii, p. 160)

Later, in continuing that passage we have twice quoted, in which Hobbes points out that authorization does not restrict a man's natural liberty, he says: 'for by allowing him [the sovereign] to *kill me*, I am not bound to kill myself when he commands me. It is one thing to say, *kill me, or my fellow, if you please*; another thing to say, *I will kill myself, or my fellow*' (*E.W.* iii, p. 204).

But no man can be supposed to authorize another to punish him, or kill him. For in authorizing, one man gives another the use of his right, but no man has the right to harm himself, for the right of nature is a right to do what is conducive to one's preservation. Fortunately, Hobbes himself recognizes this:

. . . there is a question to be answered, of much importance; which is, by what door the right or authority of punishing in any case, came in. For by that which has been said before, no man is supposed bound by covenant, not to resist violence; and consequently it cannot be intended, that he gave any right to another to lay violent hands upon his person. . . . It is manifest therefore that the right which the commonwealth, that is, he, or they that represent it, hath to punish, is not grounded on any concession, or gift of the subjects. (*E.W.* iii, p. 297)

Hobbes proceeds to give this account of the right to punish:

But I have also showed formerly, that before the institution of commonwealth, every man had a right to every thing, and to do whatsoever he thought necessary to his own preservation; subduing, hurting, or killing any man in order thereunto. And this is the foundation of that right of punishing, which is exercised in every commonwealth. For the subjects did not give the sovereign that right; but only in laying down theirs, strengthened him to use his own, as he should think fit, for the preservation of them all: . . . (*E.W.* iii, pp. 297–8)

But this account is a reversion to Hobbes's initial theory of sovereignty. Instead of treating sovereign right as artificial, the right to use the several rights of the subjects, he treats it as natural, the right of an individual who has not laid down his natural right, when all others have laid down theirs. If we must take this to be the only explanation of the right of punishment, then we have found a major weakness in Hobbes's later theory—a weakness which may lead us to suppose that sovereignty cannot be sufficiently founded on authorization.

However, we need not acquiesce in such a conclusion. For in the state of nature, as Hobbes points out, every man has the right to kill, injure, or subdue others. Now in the institution of the commonwealth, each man authorizes the sovereign to use his right of nature, and as part of that right, therefore, his right to kill, injure, or subdue others. The right to punish may then easily be explained. Each man authorizes, not his own punishment, but the punishment of every other man. The sovereign, in punishing one particular individual, does not act on the basis of his authorization from that individual, but on the basis of his authorization from all other individuals.

This does, of course, mean that each subject is not author of everything the sovereign does, although some subject is author of everything the sovereign does. It does mean that the right to punish is subtly different from all other rights of the sovereign, in that the sovereign in each act of punishing is not exercising a right given him by all of his subjects. It does mean that in punishing, the sovereign is no longer acting as the representative of the person punished, and so that he is placing himself in the position of an enemy in the state of nature with respect to that person. But of course the person punished, in violating the civil law, has violated an obligation undertaken in the institution of the sovereign, and so has already placed himself, in effect, in the state of nature with respect to the other members of civil society, as represented in the person of the sovereign.

The sovereign right to punish thus does require a complication in Hobbes's theory of authorization. But it is a complication which preserves the essential character of that theory, and so we may well allow it. Although Hobbes was unable to

explain the right to punish in a manner consistent with his
own theory, we have, I believe, succeeded where he failed.

B. *Authorization, Obligation, and Covenant*

In this section we shall first set out the doctrine of political
obligation which we have found in Hobbes's arguments. We
shall then criticize two very different views—those of Oakeshott and Warrender. In the next section we shall consider the
question whether the subjects can be said to have an obligation *to* the sovereign.

Sovereignty is founded on act of institution performed by
the prospective subjects. This act has two components, which
it is absolutely necessary to distinguish clearly. The first
component is the act of authorization, whereby the subject
confers the use of his right of nature on the sovereign. The
second is the act of covenant, whereby the subject obliges
himself to all his fellow subjects to confer the use of his right
on the sovereign, and not thereafter to withdraw the use of
his right from the sovereign.

The first component, the act of authorization itself, does
not put the subject under obligation. To confer the use of
one's right on some other person is not to deprive oneself of
that right. Hobbes makes this point evident in *A Dialogue
Between a Philosopher and a Student of The Common Laws of
England*: 'He that transferreth his power, hath deprived
himself of it: but he that committeth it to another to be exercised in his name and under him, is still in the possession of
the same power' (*E.W.* vi, p. 52).

However, we have seen that the subject *can* not exercise any
right incompatible with his authorization, *as long as* he maintains the authorization. This we must remember is a purely
logical point. If the subject does exercise some right incompatible with the authorization, then necessarily he withdraws
the authorization. It does not follow from this that the subject
may not, or has an obligation not to, exercise any right incompatible with his authorization. This requires the additional

premiss that the subject may not, or has an obligation not to, withdraw his authorization.

But this additional premiss is provided by the second component, the act of covenant. By covenant the subject has given up his right to withdraw his authorization from the sovereign. From this it follows necessarily that the subject has given up all those rights the exercise of which would be incompatible with his authorization.

Summarily, the argument is as follows:

1. If A authorizes B, and if A then performs some act incompatible with his authorization of B, then A thereby withdraws his authorization from B.

2. Therefore if A authorizes B, and if A has an obligation not to withdraw his authorization from B, then A has an obligation to perform no act incompatible with his authorization of B.

3. But if A is a subject, then A authorizes his sovereign, and A has an obligation not to withdraw his authorization from his sovereign.

4. Therefore if A is a subject, then A has an obligation to perform no act incompatible with his authorization of his sovereign.

This obligation we term *political obligation*. We can determine its extent by showing what acts are incompatible with the authorization of the sovereign. And this of course should correspond with the extent of those acts which do not fall under our material definition of the liberty of subjects.

We shall take for granted throughout that the scope of political obligation does not extend to acts directly inimical to the preservation of the agent. That every man has an inalienable natural right to do what is necessary for his own preservation is a general limitation on all obligation. Hence we need not specify this limitation in each particular case of obligation.

Now it is evident that political obligation must include both an obligation not to oppose the sovereign, and an obligation to obey the sovereign when obedience is necessary to either internal or external security. To oppose the sovereign would be to refuse to recognize his acts as one's own, and this would be to withdraw one's authorization from him. This obligation corresponds directly to that obligation which, according to Hobbes's initial theory of sovereignty, is assumed by the subject in parting with his right to resist the sovereign.

To refuse to obey the sovereign when obedience is necessary to internal or external security would be to refuse to the sovereign the means necessary to the ends for which he was instituted. And this would be incompatible with maintaining one's authorization of the sovereign *for the purpose of* attaining those ends. It will be remembered that the covenant of authorization specifies the ends for which the sovereign was instituted, so that they are integral to the act of authorization.

So much is unproblematic. But does political obligation extend to an obligation to obey the sovereign when obedience is *not* necessary to internal or external security? According to our material definition of the liberty of subjects, a subject has no liberty to do what is contrary to the civil law except when his preservation gives him sufficient ground. But can we demonstrate an obligation to conform to civil law—that is, obey the sovereign—if the ends of the institution of sovereignty will not be jeopardized by disobedience?

Hobbes specifically exempts subjects from such an obligation if obedience would be dangerous to them. He argues:

> ... the obligation a man may sometimes have, upon the command of the sovereign to execute any dangerous, or dishonourable office, dependeth not on the words of our submission; but on the intention, which is to be understood by the end thereof. When therefore our refusal to obey, frustrates the end for which the sovereignty was ordained; then there is no liberty to refuse: otherwise there is. (*E.W.* iii, pp. 204–5)

But this exemption simply corresponds to the second part of our material definition of the liberty of subjects. It leaves aside the case in which obedience is neither necessary to the society, nor dangerous to the subject. Has the subject then an obligation to obey?

We may appeal here to Hobbes's statement that '(*t*)*he sovereign is judge of what is necessary for the peace and defence of his subjects'* (*E.W.* iii, pp. 163–4). The subjects have an obligation to accept the sovereign's judgement of what means are necessary to the ends for which he was instituted, since their refusal to accept his judgement would make it impossible for him to attain those ends. Therefore, the sovereign may rule that obedience to his commands is always necessary to the maintenance of society; the subjects are obliged to accept his judgement, and so obliged to obey his commands. The question whether obedience actually is necessary does not arise.

But if the question does not arise, then the subjects can never take advantage of the exemption Hobbes offers them, when obedience is not necessary to society and is dangerous to the subject. The subjects will have only the inalienable exemption from obligation, when obedience would be directly inimical to their preservation. For this exemption does not depend on the judgement that obedience is not necessary to society—a man has the right to preserve himself, whatever the consequences to the commonwealth.

Let us suppose the question does arise. Or we may suppose that the sovereign does not rule that obedience to his commands is always necessary to the ends for which he was instituted; he leaves room for the judgement of the subject. Once again, then, has the subject an obligation to obey?

It seems possible to argue that, although obedience to some of the commands of the sovereign is not necessary to the attainment of the ends of the commonwealth, such obedience is in almost all cases conducive to these ends. That sovereign who is more fully obeyed by his subjects is better able to provide for internal and external security. Therefore it would be incompatible with the authorization given the sovereign, for the subject to refuse obedience in those cases when obedience would not be dangerous to him.

Thus political obligation includes:

1. An obligation not to oppose the sovereign.
2. An obligation to obey the sovereign, when obedience is necessary to internal or external security.

3. An obligation to obey the sovereign, even if obedience is not strictly necessary to internal or external security, as long as obedience is not dangerous.

These obligations, taken with the general limitation stated at the outset, provide a sufficiently close counterpart to the liberty of subjects, as stated in our material definition. But we have been forced to extend Hobbes's own arguments somewhat—though not to violate the principles of his theory—in order to include the third of these obligations.[1]

The first of the two alternatives to our account of political obligation which we propose to consider is that set forth by Michael Oakeshott in his well-known Introduction to *Leviathan*. The crucial passage in Oakeshott's argument is this:

> An authority is a will that has been given a Right by a process called authorization, which . . . is the voluntary act of those who are to be morally obliged or bound by the commands of the authorized will. This voluntary act of authorization is a surrender . . . of the natural Right of each man, which, in a single act, creates and endows with authority an artificial Representative . . . who . . . is called Sovereign. The exercise of the will of the Sovereign is called legislation, and moral obligation is the offspring of laws so made.[2]

Now one of the most striking features of this argument is the lack of textual support Oakeshott finds—or that can be found—for it. But we need not rely on a merely negative

---

[1] I offer the following general test, which the subject may apply to any proposed action, in order to determine his political liberties and obligations with respect to it:

1. Is the action (a) necessary for or (b) incompatible with his preservation? If (a), reason dictates its performance, and he is at liberty to perform it. If (b), reason dictates its omission, and he is at liberty to omit it. If neither, he proceeds to

2. Is the action (a) necessary for or (b) incompatible with the maintenance of his authorization of the sovereign? If (a) he has an obligation to perform it; if (b) he has an obligation to omit it. If neither, he proceeds to

3. Is the action dangerous (a) to omit or (b) to perform? If (a) he is at liberty to perform it; if (b) he is at liberty to omit it. If neither, he proceeds to

4. Is the action (a) commanded by or (b) forbidden by the sovereign? If (a) he has an obligation to perform it; if (b) he has an obligation to omit it. If neither, he is at liberty to perform or to omit it.

[2] M. Oakeshott, op. cit., p. lx.

objection. For we may easily show that this is not Hobbes's view of political obligation.

What is crucial in Oakeshott's argument is the claim that authorization 'is a surrender . . . of the natural Right of each man'. But authorization, we have shown, is the procedure by which one man gives the use of his right. This in no way involves the surrender of the author's right to the actor. To suppose the author to surrender his right would be incompatible, both with Hobbes's insistence that authorization is operative in non-political contexts in which one man licenses another to act on his behalf, and with those explicit passages, which we have already quoted, in which Hobbes insists that authorizing the actions of another does not involve restricting one's liberty or giving up one's power (*E.W.* iii, p. 204; *E.W.* vi, p. 52).

But perhaps Oakeshott supposes that the authorization of the sovereign is a very special case of authorization, in which the prospective subject gives to the sovereign the right to govern him. There is some textual support for this: Hobbes does represent the authorization in these words: '*I authorise and give up my right of governing myself, to this man, . . .*' (*E.W.* iii, p. 158).

But what is this 'right of governing myself'? It cannot be simply the right to perform acts owned by me; this I do give in authorization, but I am not therefore obliged to obey the person to whom I give this right. Nor can it be the right to issue commands which I am obliged to obey, for it makes no sense to suppose that I possess such a right. Such a right would require that I be able to oblige myself *to* myself, but Hobbes says 'nor is it possible for any person to be bound to himself; because he that can bind, can release; and therefore he that is bound to himself only, is not bound' (*E.W.* iii, p. 252).

The 'right of governing myself' must, it seems, be the right to make decisions which are to be counted as mine. But although I can authorize another man to make such decisions, I am not therefore obliged to act on them, for I am not obliged to act on my own decisions. I am at liberty to change my mind, and decide otherwise. Of course, if I authorize someone to make decisions on my behalf, and which determine what I am to do, then refusal to act on them is tantamount to withdrawing my authorization from him. But I am not obliged to maintain my authorization.

However, it may be urged, we have not attended to the force of 'up'—I 'give up my right of governing myself'. I give up the right to make decisions which are to be counted as mine. Hence I am obliged to act on such decisions, since not to act on them is to make other decisions, and I have given up my right to do so. And I may not withdraw my authorization, and thus resume the right to make decisions, since to withdraw my authorization requires a decision which I have given up the right to make.

This is all very well. But Hobbes never characterizes authorization as a procedure for giving *up* right. His justification for speaking in this way in the passage quoted is, of course, that every man covenants to give his right of governing himself to another, and so, being obliged by his covenant to do nothing incompatible with his authorization, can be said to have given *up* his right. But then obligation cannot be made to depend on the process of authorization; only a prior obligation to keep one's covenants will explain why the subject has given up his right to govern himself and so must regard the laws of the sovereign as his own decisions which he has no right to alter.

Oakeshott's account of obligation to the laws of the sovereign, then, simply obscures the logic of Hobbes's argument for political obligation. Hobbes never relates his concepts of obligation and authorization in any way which would make Oakeshott's account intelligible.

Howard Warrender presents a very different account of political obligation. His argument may, I think, be fairly summarized in this manner:

(i) '(N)othing can be transferred by the political covenant except rights of resistance'.[1] Therefore the covenant itself serves

---

[1] Howard Warrender, op. cit., p. 113.

only to introduce an obligation not to resist the sovereign. It is this obligation which Hobbes supposes to arise from the *words* of the covenant.[1]

(ii) But Hobbes also supposes that the citizen has 'a general obligation to act in such a way that the purpose of the covenant shall not be frustrated'.[2] This obligation cannot be introduced by the covenant itself. Hobbes supposes it to arise from the *intention* of the covenant, which 'is to be understood by its purpose. . . . He [Hobbes] takes as *the purpose* of the covenant the maintenance of a political society, and the citizen is regarded as having a positive duty to secure this end.'[3]

(iii) In appealing to the intention of the covenant, in order to justify the positive duties of the subject, Hobbes is attempting 'to give these cases the appearance of falling under the principle of keeping covenants', but this and his appeal in places to tacit covenant 'are thin disguises for different principles'.[4]

(iv) Therefore, 'an interpretation of his doctrine that represents political obligation as the simple effect of the obligatory character of a covenant, cannot accommodate its scope'.[5]

(v) 'The use he [Hobbes] makes of an obligation to further the objective of the covenant, peace and society' relies 'upon a general obligation to observe the laws of nature'.[6]

Thus Warrender's account of political obligation serves to support strongly his thesis that the laws of nature must be considered as laws, and as obligatory. If not, then on Warrender's view no adequate basis can be provided for political obligation.

But it will be readily seen that Warrender's argument rests on a misconception about the role of authorization. He fails to realize that in authorization Hobbes has an entirely new device for 'transferring' right. It is true that in his initial theory of sovereignty nothing can be transferred by the political covenant except the right of resistance, since the only

[1] Howard Warrender, op. cit., p. 190.   [2] Ibid., p. 113.   [3] Ibid., p. 235.
[4] Ibid., p. 236.   [5] Ibid., p. 114.   [6] Ibid.

form of transfer possible is to renounce one's rights in favour of the sovereign. But in *Leviathan* what is transferred is the use of one's right.

The covenant then introduces, not an obligation of non-resistance to the sovereign, but an obligation to maintain one's authorization of the sovereign. And it depends strictly on the words, and not the intention, of the covenant. For the covenant, stripped to its essentials, runs 'I authorize A, to the end that he may provide internal and external security, provided you do so as well'. One's obligation is to carry out this covenant—that is, to authorize the sovereign, and maintain one's authorization, for the end specified. This obligation is all that is required by Hobbes's political theory, as we have shown.

The distinction Hobbes makes between the words and the intention concerns the act of authorization and not the act of covenant. The content of political obligation clearly depends on the intention of the authorization, and not on the mere words 'I authorize all his actions' which, as we have seen, impose no obligation at all. But Warrender, failing to distinguish the act of authorization from the act of covenant, supposes that Hobbes thinks that political obligation depends on the intention of the covenant, and so cannot be 'the simple effect of the obligatory character of a covenant'.

Warrender's argument, then, never gets properly started. And since Hobbes does not appeal to the intention of the covenant in order to justify political obligation, steps (iii), (iv), and (v) collapse utterly. And so we have no reason to suppose that Hobbes must appeal to a general obligation to obey the laws of nature in order to account for political obligation. Indeed, we have instead every reason to suppose that only an obligation to keep one's covenant is needed, and this obligation we have shown to be quite independent of any supposed obligation to obey the laws of nature.

C. *Obligation to the Sovereign*

Political obligation, as we have characterized it, is owed by each of those who covenant to authorize the sovereign to every other covenanter—that is, by each subject to every other subject. It seems to follow, therefore, that political obligation could be terminated by each subject releasing every other subject from the covenant. And this would be, in effect, to depose the sovereign.

Hobbes will not admit this possibility. He claims therefore that the subjects are obliged, not only to each other, but also to the sovereign. We saw that the arguments used to support this claim in the initial theory of sovereignty are unsatisfactory. Does Hobbes do better in *Leviathan*?

Hobbes reasons in this manner:

> . . . they have also every man given the sovereignty to him that beareth their person; and therefore if they depose him, they take from him that which is his own, and so again it is injustice. Besides, if he that attempteth to depose his sovereign, be killed, or punished by him for such attempt, he is author of his own punishment, . . .: and because it is injustice for a man to do anything, for which he may be punished by his own authority, he is also upon that title, unjust. (*E.W.* iii, p. 160)

We have already disposed of the second argument here, in showing that Hobbes cannot consistently suppose a man ever to be author of his own punishment. And the first argument fares no better. For it is not unjust to withdraw one's authorization from someone whom one has previously authorized. If it were, a man might never dismiss those whom he commissions to represent him, which is absurd. If, then, it is unjust to dismiss the sovereign, because it would be to take from him that which is his own, the sovereign must possess some right to represent his subjects beyond that which they give him in authorizing him. But Hobbes never suggests that the sovereign possesses such an additional right.

It might be urged that the injustice lies, not in depriving the sovereign of some right beyond that conferred in authorization, but in taking from him what has been freely given him. For it is injustice to take back one's free gift. But the sovereign has not been freely given the use of his subjects' rights; he has been authorized to use these rights, and authorization is not gift.

Hobbes, then, has no argument to support his claim that the subjects have an obligation to the sovereign, from which they may not release themselves. However, it is possible to come to his rescue here. Although no argument will show that the subjects have an obligation to the sovereign, it is possible to show that the subjects may not release themselves from their obligation, to each other, to maintain their authorization of the sovereign. And this is sufficient for the purposes of Hobbes's theory.

I propose now to construct a Hobbesian proof that the subjects may not release themselves from their political obligations. This proof is not given, not even hinted at, by Hobbes. But I believe that it makes use only of premises which may be found explicitly in his system, or which he uses in his own arguments.

We may think of the right, possessed by every man in the state of nature, to enter into covenants, as tripartite. Each man has a right to offer or propose covenant, a right to accept covenant, and a right to release from covenant. The second of these rights may be called the right *to bind*; if A accepts B's offer of covenant, then A binds B to performance of the covenant. B now has an obligation owed to A to perform. If A later exercises the third right, the right to release, then A releases B from performance of the covenant. B's obligation, owed to A, is thereby terminated.

As we have seen, Hobbes says that 'he that can bind, can release' (*E.W.* iii, p. 252). But it does not follow from this that the two rights, or their use, are never separable. Suppose A authorizes B to act for him in purchasing a house. B now has the use of A's right to offer covenant, and the use of A's right to bind, in the matter of purchasing a house. But B does not therefore have the use of A's right to release. B may bind C to sell his house to A for £15,000, but B may not release C from this obligation, unless he has separate authorization from A to do so.

This is a simple and obvious case in which the right to bind is separable from the right to release. I introduce it simply to show the possibility. Now let us consider a quite different and more curious case. Suppose A and B make a covenant, one of the terms of which is that each forgoes the right to release the other from the covenant. This could not be the complete content of the covenant, of course, but there is no reason to suppose that it could not be part of the content.

A then has an obligation not to release B, and B has a similar obligation not to release A. Neither then may release the other. For A to exercise the right to release B, B would first have to release A from his obligation not to release B. But for B to exercise the right to release A from this obligation, A would first have to exercise his right to release B from his obligation not to release A. Thus release is made impossible.

Note that each must forgo the right to release the other. If A alone renounced his right to release B, but B did not renounce his right to release A, then A could exercise the right to release B if B first released A from his obligation not to release B, which B would retain the right to do. A unilateral obligation not to release the other party would thus be pointless; only a mutual obligation can be effective.

Now we ask whether the political covenant obliges each of the covenanters not to release the others. It does, if it would be incompatible with the intention of the authorization of the sovereign for the subjects to retain the right to release each other from the covenant of authorization. But this condition is met. For if the sovereign must act under the continual threat of deposition by his subjects releasing each other from their covenant to authorize him, then his efforts to fulfil the intention of the authorization, by guaranteeing the internal and external security of the commonwealth, will be jeopardized.

Therefore, in covenanting to authorize the sovereign, the subjects put themselves under obligation to each other, not to release each other from their covenant. And this, as we have seen, is to make release impossible—at least in so far as it depends upon acts which the subjects have the right to perform.

The sovereign, of course, in receiving the use of the rights of his subjects, receives also the use of the right to release his subjects from their covenant of authorization. That is, the sovereign may exercise the right of subject A to release B, C, and the rest, of subject B to release A, C, and the rest, and so on. But this is quite innocuous, for the sovereign in any case has the right to renounce his own position. As long as the sovereign need fear only deposition by himself, his position is as secure as he can want. His effectiveness in fulfilling the intention of the authorization is in no way jeopardized.

The subjects are released from their covenant of authorization of the sovereign if the sovereign fails to provide them with security. But this does not require that the subjects exercise any right to release each other. As Hobbes insists, '(t)he obligation of subjects to the sovereign, is understood to last as long, and no longer, than the power lasteth, by which he is able to protect them' (*E.W.* iii, p. 208). For 'the obligation of subjects to the sovereign' we must read 'the obligation of subjects to maintain their authorization of the sovereign', but this makes no difference in this argument. For release from the covenant of authorization follows automatically if the inalienable right of nature, possessed by each subject to defend himself, requires this release. And this is required if the sovereign fails to protect the subjects.

I conclude that Hobbes is correct in maintaining that the subjects may not depose the sovereign, though his arguments are mistaken. That it should be possible to construct a better argument than Hobbes himself gives is not surprising; it is no discredit to Hobbes that he should not have developed all of the consequences of his theory. Indeed, I should look upon the construction of this argument as a tribute to the power and coherence of Hobbes's theory of sovereignty.

4. SUCCESS OR FAILURE?

A. *Sovereignty and Security*

Hobbes's great aim is to show men the way to security. His argument is that only in a commonwealth ruled by a sovereign who exercises without limitation the rights of all his subjects, and whose power is sufficient to make that exercise effective, can men find security. But can there be such a sovereign? And does he provide security for his subjects? These questions we must now attempt to answer.

We shall examine three major problems in Hobbes's account of sovereignty. First, we shall suppose that it is possible for commonwealth, as Hobbes describes it, to exist, and we shall ask whether the state of civil society is truly preferable to the state of nature. Then we shall question our supposition. Our second problem will be whether it is possible for men to give to one man or assembly the right and power which Hobbes deems necessary. And our third problem will be whether this right and power, if given, can actually be exercised.

John Locke argues that 'absolute monarchy . . . is indeed inconsistent with civil society'.[1] Civil society depends on 'a known authority to which every one of that society may appeal upon any injury received or controversy that may arise', but the subjects of an absolute monarch can appeal to no such authority in controversies with him. And 'wherever any persons are who have not such an authority to appeal to and decide any difference between them there, those persons are still in the state of nature'. The subjects remain, then, in the state of nature, with respect to the sovereign.

This Locke considers absurd. Men seek to escape the state of nature, not to compound it. Although Locke does not refer to Hobbes by name, it is difficult to doubt that he refers implicitly to Hobbes's doctrines, when he writes:

As if when men quitting the state of nature entered into society, they agreed that all of them but one should be under the restraint of laws, but that he should still retain all the liberty of the state of nature, increased with power, and made licentious by impunity. This is to think that men are so foolish that they take care to avoid what mischiefs may be done them by polecats or foxes, but are content, nay, think it safety, to be devoured by lions.[2]

---

[1] John Locke, op. cit., section 90.
[2] Ibid., section 93.

Locke's argument is not quite fair. There is only one lion, and one lion may be less hurtful than a horde of polecats or foxes. Furthermore, although Locke's account fits Hobbes's initial theory of sovereignty, it does not fit the later theory— the sovereign does not retain the liberty of the state of nature, but as we have seen acts as the representative of his subjects.

Formally, then, Locke's objection is mistaken. Locke seems to presuppose that the relations between man and man fall into two exhaustive and exclusive classes—those between men not subject to a common authority, and those between men subject to such an authority. Men related in the first way are in the state of nature with respect to each other; men related in the second way are in civil society. But Hobbes provides a third class of relationships—those between an author and an actor. And in this third class falls the relation between subject and sovereign—a relation, then, which is neither natural nor social, but the basis for converting natural relationships into social relationships.

But to object to Locke's argument on merely formal grounds is to ignore its very real point. Hobbesian society, this objection runs, is worse than the state of nature—worse, because in the state of nature each man is at least contending only with his equals, whereas in civil society each man is contending with someone who, although styled his representative, is actually supreme over him both in right and power. And this supremacy, we can be sure, will be exercised not in the interests of the nominal authors of sovereign right, but in the interests of the actor, the sovereign himself.

The elaborate doctrine of authorization, with its consequences for sovereign right, and the liberty and obligation of subjects, is a mere façade. The key to the façade is the incorporation of despotic sovereignty into the doctrine. The covenant offered by the vanquished to the victor, 'that so long as his life, and the liberty of his body is allowed him, the victor shall have the use thereof, at his pleasure' (*E.W.* iii, p. 189), is in fact a covenant 'of owning, and authorizing whatsoever the master shall do' (*E.W.* iii, p. 190).

But if the vanquished authorizes the victor, and the servant authorizes the master, then authorization is no more than a convenient device whereby power can assure itself of legitimacy, of right. The sovereign scrupulously acts as the representative of his subjects, because they cannot but choose to give him the right to represent them. He respects their liberties, and requires only that they fulfil their obligations, because he determines the content of those liberties and obligations. For the subjects to object, is for them to invite death.

Hobbes says, 'The *people* rules in all governments' (*E.W.* ii, p. 158). The people rules, because 'the king is the *people*' (ibid.). The Hobbesian sovereign is in the position of the Bonapartist Emperor or the Nazi Führer. All Hobbesian governments are, in effect, plebiscitary democracies in which 'the *people* wills by the will of *one man*' (ibid.). We can reasonably suppose that the up-to-date Hobbesian sovereign would not rest his position on an assumed act of authorization, but would demand, and of course receive, repeated plebiscitary endorsements of his right to rule. If the sovereign is iniquitous, God may punish him, but the subjects can complain of no injustice.

If all this is allowed, Hobbes could insist that it constitutes no objection to his argument. For the evils of civil society must be set against the evils of the state of nature. And if the state of nature is truly intolerable, then civil society must be preferred. The question is not whether civil society is unpleasant, but whether it is less pleasant than some possible alternative. Thus the only effective objection must be to show either that the state of nature is not a state of war of all against all, or that the rights and powers of the sovereign can be limited without sacrificing the advantages of civil society. And both of these, Hobbes could claim, are ruled out by the nature of man. If there is an objection it is not to the political theory, but to the psychology.

But Hobbes need not even allow all that this criticism contains. If it were a fair representation of his position, then his own presentation of his political theory would be disingenuous. But it is not. It is a statement of an extreme situation which Hobbes would have to allow to be possible, and against which Hobbes could provide no theoretical objections. But neither reason nor passion need move a sovereign to become an arbitrary despot. Some men 'would be glad to be at ease within modest bounds' (*E.W.* iii, p. 112) if they could be secure; if some men, then why not some sovereigns, who can be secure?

But if civil society is preferable to the state of nature, is civil society possible? In the preceding chapter we considered the evident problems in Hobbes's account of how the sovereign can be supposed to acquire power commensurate with his right. These problems are conveniently obscured in *Leviathan*; Hobbes no longer vacillates between a transfer of right and a transfer of power, but boldly insists that the subjects 'confer all their power and strength upon one man' (*E.W.* iii, p. 157). And referring to the act of authorization he claims that 'by this authority, given him by every particular man in the commonwealth, he [the sovereign] hath the use of so much power and strength conferred on him, that by terror thereof, he is enabled to perform the wills of them all, to peace at home, and mutual aid against their enemies abroad' (*E.W.* iii, p. 158).

But Hobbes's failure to state once again that power cannot literally be transferred should not be taken to indicate that he has altered his view. For it is evident that no act concerning right—whether it be renunciation or authorization—can literally effect a transfer of power. A man can promise the use of his power to another, but he cannot literally give him that use. Hence the account of the institution of sovereignty in *Leviathan* offers no solution to the problem of establishing sovereign power.

The account in *Leviathan* is nevertheless an improvement on Hobbes's earlier statements. For it suggests immediately what is required if the power of the sovereign is to be commensurate with his right. Sovereign right is the use of all of the rights of the subjects; similarly, then, sovereign power is the use of all their powers. Sovereign power cannot be given to the sovereign; it cannot be exercised directly and only at his will. But sovereign power can be allowed the sovereign; the subjects can conform their wills to his. Thus what is required of every man, if sovereignty is to be effective, is that he own all of the sovereign's actions, not merely in acknowledging them as his own, but in adding his will to them, so that his own actions conform to the conduct of the sovereign.

But there is a difficulty here. A man cannot determine his own will; he cannot will to will, as Hobbes insists (*E.W.* iv, p. 69). Thus he cannot will to add his will to the sovereign's actions. But it is not necessary that he do this. All that is required is that he be sufficiently motivated, so that he does will to conform his actions to the will of the sovereign. And thus sufficient motivation is found, when his own particular interests are not in conflict, in his general desire for peace and security.

The sovereign can thus rely on the general support he may expect from his subjects to coerce recalcitrant individuals whose interests are adversely affected by particular acts which he performs. In this way the sovereign consolidates the power necessary for his right to be effective. As we saw in the preceding chapter, no greater power than this can be established by men. It is the best they can do to make themselves secure.

There is, however, a further problem, which concerns the exercise of sovereign right and power. Hobbes insists, not only that sovereign power be unlimited, but also that it be indivisible. 'For what is it to divide the power of a commonwealth, but to dissolve it; for powers divided mutually destroy each other' (*E.W.* iii, p. 313). This doctrine has for Hobbes a very practical significance, for '(i)f there had not first been an opinion received of the greatest part of England, that these powers were divided between the King, and the Lords, and the House of Commons, the people had never been divided and fallen into this civil war' (*E.W.* iii, p. 168). How then is the sovereign alone to rule over the multifarious aspects and affairs of a commonwealth?

Hobbes recognizes that in fact the sovereign alone cannot do so. He requires counsellors, magistrates, ministers, and other agents whom he authorizes to act on his behalf (cf. chapter 23 of *Leviathan*). Without these the sovereign would be isolated from society, unaware of the problems requiring his decision, and unable to translate his decision into an effective resolution of the problems. These agents of the sovereign power must then in practice, and contrary to Hobbes's theory, share the exercise of that power with the sovereign. The sovereign can only be the focal point of a set of rights and powers whose actual exercise is divided among all those who represent him.

Behind the façade of the unity of the sovereign is the reality of a division of powers. Hence if Hobbes insists that only unlimited and indivisible power is sufficient to maintain human society, he must conclude that society is impossible. If the formal division of sovereignty is ruled out, then the informal division among the sovereign's agents is also ruled out. They will compete for power among themselves, destroying each other and the sovereign in the process. Hence civil society, far from being the stable and secure condition Hobbes envisaged, is at best a truce among those who contend for the reality of power behind the façade of undivided sovereignty—a truce which must be broken, with a resultant war which ends in a return to the state of nature.

If we turn from Hobbes's arguments to reflect briefly on the concentration of power, as we know it in our own civil societies, several significant points come immediately to our attention. Any actual concentration of power depends upon the effectiveness of institutional arrangements which provide methods for making decisions concerning large numbers of men, and machinery for carrying out and enforcing these decisions. The actual working of these institutions is not relevant to our purpose; we need note only that they must exist.

First, then, it is evident that both the nature and the extent of the powers concentrated are functions of these institutional arrangements. Hence Hobbes is quite wrong to suppose that he who would limit sovereign power must invoke a greater power to impose the limitation (cf. *E.W.* iii, p. 195). Limitation of sovereign power is inevitable; we may choose among different institutional arrangements for the differing concentrations of power they provide, but no arrangement provides total concentration. And Hobbes is not in fact unaware of this; 'there is no commonwealth in the world, wherein there be rules enough set down, for the regulating of all the actions, and words of men; as being a thing impossible' (*E.W.* iii, p. 199). Hence his objection to the possibility of limiting sovereignty rests on a simple failure to consider sufficiently what in fact he recognizes.

Second, as a consequence of this no society could effectively regulate the behaviour of men if all men were concerned to exploit every weakness, every lack of power to enforce obedience, to increase their own strength at the expense of their fellows. Not all men need be tractable, and no man need always be tractable, but unless most of the men most of the time are prepared to accept social regulations and arrangements, not because of an immediate compulsion to obedience, but because of a customary disposition to obedience, then political society as we know it becomes impossible.

Third, tractability is required not only among the general body of subjects but especially among the officials authorized by the sovereign. For the institutional arrangements which make any concentration of power possible depend for their effectiveness on the willingness of those who staff the institutions to co-operate in maintaining them, rather than only to compete in aggrandizing power. This co-operation cannot rest entirely on the fear of some further power. At some point, barring a Kafkaesque regress without limit, some degree of voluntary co-operation on the part of some persons is the condition of the continued working of all social and political bodies.

To claim that tractability is required among those who exercise ruling power and among those who are ruled is not to suggest that all government rests on co-operation and consent. Some co-operation among those who are to exercise power is required if power is to be concentrated; some consent from those over whom power is to be exercised is required if this power is to be effective. But both the co-operation and consent may be so limited that the labels prove misleading. A government which relies primarily upon fear in order to train both rulers and ruled to acquiesce in the order it provides is itself the primary source of the acquiescence it requires in order to continue to exist.

If Hobbes's gloomier thoughts were sound, then the tractability required for any political and social order among men would not exist. But the existence of such orders is proof that Hobbes overstates his case. And if we remove the overstatement, allowing men even that minimum willingness to co-operate freely with their fellows in maintaining and respecting institutions for the concentration of power, then Hobbes's awful choice—between the state of nature and unlimited sovereignty—disappears.

We need not avoid anarchy by acquiescing in what, despite Hobbes's dislike of the term, we may call tyranny.[1] Both are extremes—anarchy the extreme condition in which no power is concentrated, tyranny the extreme condition in which all power is concentrated. They limit, but do not represent, the actual condition of man. Between them, within the limits set by man's actual tractability and by the existent threats to human survival, we may choose. Our choice may of course be unwise. Our institutions may permit the concentration of power, not in the interests of the security and well-being of the community, but primarily in the interests of those who occupy positions of power within the institutions. Such a concentration will approach tyranny. Or our institutions may permit a concentration of power insufficient to hold men in any check—insufficient police power to prevent internal lawlessness, or

---

[1] 'For they that are discontented under *monarchy*, call it *tyranny*; . . .' (*E.W.* iii, p. 171). But Hobbesian men are always discontented. We extend the term, in the modern fashion, to include non-monarchical governments.

insufficient military power to withstand external hostility. Such a concentration will approach anarchy.

Civil society may fail to provide men with permanent security, and with the liberty to promote their own ends. But Hobbes, who recognized the uncertainty and impermanence of all human affairs, can hardly complain if the commonwealth cannot guarantee the benefits which it does often provide. To assert that man can attain only an insecure security is neither to deny that man does often attain security nor to assert that he must always remain in insecurity.

We may draw a twofold conclusion from our discussion of this third problem. Given Hobbes's account of man, and his view of how power may be concentrated, he is not really able to offer any alternative to anarchy—not even what his critics consider tyranny. When we examine the internal structure of the mighty Leviathan, we find that on Hobbes's premisses it will not work. The sovereignty necessary for security is not attainable.

But if we admit a different view of man, which includes recognition of human tractability, and if we understand the importance of institutional arrangements both in creating and in limiting concentrations of power, then the problem of security takes on a quite different appearance. Hobbes's simple alternatives—the anarchy of the state of nature and the tyranny of the absolute sovereign—fade into ideal limits. The real problem is to determine that relationship between institutions and power which provides sovereign power sufficient to make us secure in the exercise of our liberties, without being sufficient to determine of itself the extent of those liberties.

B. *The Metaphor of Authorization*

The outstanding merit of Hobbes's political theory is not to be found in its content. The impressive logic with which Hobbes moves from his premisses to the conclusion that absolute sovereignty is necessary to human security, cannot finally conceal either the inadequacy of the premisses or the failure to show that absolute sovereignty is possible. Reason enough, one might suppose, to dismiss the theory as a magnificent failure. But this is to fail to appreciate what remains, after the content is rejected—the form, the conceptual structure itself. The concept of authorization is Hobbes's enduring contribution to political thought.

Political inquiry is one of those areas of human reasoning which invites domination by a metaphor. The conventional metaphor which seems to provide the framework of Hobbes's theory, as well as Locke's, Rousseau's, and a host of others in the seventeenth and eighteenth centuries, is that of contract. The failure of this metaphor to illuminate politics was recognized by Hume, and has been sufficiently discussed by later writers. What I wish to argue is that authorization, rather than covenant, is the dominant metaphor in Hobbes's political thought, and that authorization is a much more adequate and illuminating metaphor for the formulation and discussion of political relationships.

Hobbes's political theory cannot itself be freed from dependence on contract, or in Hobbes's own terminology covenant, because it is by covenant that all obligations are introduced. But we need not wed ourselves to this artificially restrictive condition. Not that we should want to abandon it entirely—some obligations, such as that of promise-keeping, are evidently introduced by an act akin to, though not identical with, the Hobbesian act of covenanting. But political obligation is not one of these.

What I wish to suggest, then, is that the primary political relationship, between sovereign and subject, or between government and citizen, may fruitfully be regarded as a metaphorical instance of the relation between actor and author. The rights of the government and its duties, the obligations of the citizen and his rights, may be derived from this relationship, given that both government and citizen have an obligation to uphold it and to conform to it in their actions.

This obligation is not founded on covenant. What we suppose rather is that both government and citizen may have an obligation resting ultimately on considerations of interest, without any intermediate appeal to the metaphor of an act which of itself introduces an obliging consideration. And these considerations of interest need not concern solely the party obliged. Unlike Hobbes we admit in our moral theory obligations which may run counter to, because they do not depend solely on, the interests of the obliged party. We suppose that the interests of others constitute a morally valid claim on the individual, a claim which may then serve to ground his obligations in such a way that he cannot appeal to an overriding right to advance his own interests without regard to the effect of his actions on others. This supposition requires us to reject Hobbes's theory of human nature, in so far as that theory maintains that all motivation is selfish, that all men are psychologically necessitated to seek only their own good.[1]

Furthermore, we may free ourselves from the connection of right and power which tends to reduce Hobbes's whole account of authorization to a mere façade. We do not suppose that power can acquire commensurate right by threatening the option: authorize or die. Although the subject may be compelled, through fear of injury or death, to obey a power to which he is not obliged by the interests of the members of the community, such obedience results from a prudential choice of evils, not from moral obligation.

We may also recognize sovereign right even when it lacks the power to ensure its proper effectiveness. Men may have an obligation to obey a certain authority, even when they disregard that obligation and the authority lacks power to exact obedience. Of course there are limits to the lack of power compatible with sovereign right. Even those whose political myopia leads them to recognize the regime of Chiang Kai-shek as the *de jure* government of all China would surely boggle at the claim that the inhabitants of mainland China have an obligation—even a suspended obligation—to obey that regime. And if the Chinese have no obligation, then the regime has no sovereign right. *De jure* recognition serves here only as a political weapon.

---

[1] Cf. my book *Practical Reasoning*, Oxford, 1963, especially chapter xii. The brief discussion of Hobbes there is not, in my view, fully satisfactory.

Rejecting both Hobbes's psychology and his account of the concentration of power, we suppose that authorization, and so sovereign right, may be limited. Thus we take up the option never exercised by Hobbes, but suggested in his initial account of how any association whatsoever is formed:

> A multitude of men, are made *one* person, when they are by one man, or one person, represented; so that it be done with the consent of every one of that multitude in particular . . . every man giving their common representer, authority from himself in particular; and owning all the actions the representer doth, in case they give him authority without stint: otherwise, when they limit him in what, and how far he shall represent them, none of them owneth more than they gave him commission to act. (*E.W.* iii, p. 151)

Thus in summary, we depart from Hobbes in four respects:

1. Authorization of political authority, and so sovereign right, may be limited.
2. The liberties and obligations of the subject do not depend on covenant, but on the interests of the members of the commonwealth (including himself); we might of course say, with Hobbes, that they depend on the intention of the act of authorization, departing from him in our view of what that intention is and what it requires.
3. The sovereign, as well as the subjects, has certain obligations.
4. Power cannot convert itself into right by the threat: authorize or die.

The problem of determining the extent of sovereign *right*, which I take to be the primary problem of political *thought*, and the problem of providing sovereign *power* correlative to this right, which I take to be the primary problem of political *practice*, are not our concern here. The contribution of our study of Hobbes is only to suggest a metaphor within which certain aspects of these problems can be illuminatingly expressed. The metaphor itself contains no diagnosis of these problems, nor alas does Hobbes's own use of it advance their solution.

What the metaphor most obviously clarifies is the role of
government as the agent or representative of its citizens or
subjects. This feature, which is suggested directly by the
metaphor, has been almost totally obscured in conventional
accounts of Hobbes's political theory, in which the absolute
role assigned the sovereign has been allowed to obscure the
logical structure of the argument.

The sovereign exercises the right of its subjects, in their
behalf, in so far as they are unable to act effectively as individuals. In this sense the sovereign *is* the people. Hobbes's
recognition of this is prior to his actual development of
the concept of authorization; we have already quoted from
Hobbes's statement in *De Cive*: 'The *people* rules in all governments. For even in *monarchies* the *people* commands; for the
people wills by the will of *one man*; but the multitude are
citizens, that is to say, subjects. . . . And in a *monarchy*, the
subjects are the *multitude*, and (however it seem a paradox)
the king is the *people*' (*E.W.* ii, p. 158).

As long as the sovereign acts within the bounds of its authorization—which for Hobbes are the bounds of the law of nature
—the citizens are responsible for its acts. Hobbes's account
permits us to understand the sense in which collective responsibility may be applied to the members of a political society.
Here his doctrine, firmly founded on individualism, avoids
the false inference that no man can be responsible except for
what he does himself. In allowing another to act in my name,
I allow myself to be held to account for what he thereby does.
So to commit myself is, for Hobbes, to sign a blank cheque.
In our view it is rather to sign a cheque containing only
the restriction 'Not to exceed . . .'.

The identification of the sovereign with the people does not
exclude all distinction between them. The sovereign is the
people—this is one side of an important truth. The sovereign
is over the people—this is the other side, equally emphasized
by Hobbes in his insistence on the double signification of the
word 'people'. 'In one sense it signifieth only a number of
men, distinguished by the place of their habitation; as the

people of England, . . . In another sense, it signifieth a person civil, . . . in the will whereof, is included and involved the will of every one in particular' (*E.W.* iv, pp. 145–6).

In exercising the right assigned it by the citizens, the sovereign may command their obedience as subjects. Bound to maintain their authorization, the citizens must recognize that their representative is also their commander. The subject is, as it were, the instrument of his instrument: he must act on behalf of the sovereign in executing its commands, so that it may act on behalf of him in preserving his life and liberty. It is this dual relation which necessarily occasions the problem of political authority—how to ensure that neither relationship obscures the other, that neither half of the political truth becomes the whole.

The citizen retains also the right to act on his own behalf, in so far as the exercise of this right is compatible with his authorization of the sovereign. This right necessarily exists; as we have seen, even Hobbes recognizes that the sovereign cannot regulate all matters (*E.W.* iii, p. 199). But it may be greater or smaller in extent; once again the problem is to ensure that this right or liberty neither hinders the effectiveness of the sovereign in exercising his authorized right, nor is hampered by the power of the sovereign exercised without right.

Both as subject, and as free agent, the citizen is a person distinct from the person of society. The people, as a multitude of subjects, is but the agent of the people, as sovereign. The people, as a multitude of free agents, is but a group of individuals acting outside the right and power of the people, as sovereign.

Just as the sovereign both is and is not the people, so the people both is and is not the sovereign. The sovereign, as an individual man or an assembly, may be distinguished from the people. The sovereign *per se* is an artificial person, the product of the authorizations of the several citizens. But this artificial person is a real man or group of men, and this man or group has its own individuality or character apart from the sovereignty conferred upon it. An individual who is sovereign may confound his own natural right with the authorized rights he is to exercise, and so may will and act on his own behalf rather than on behalf of the citizens. Equally, an assembly which is sovereign may have regard to the interests of its members, rather than to those of the society as a whole.

Hobbes's awareness of this problem led him to recommend monarchy as that form of government in which the interests of the bearer of sovereignty are most nearly identified with the interests he is authorized to promote (*E.W.* iii, pp. 173–4). Whether he is right in this is, however, of no concern here.

The metaphor of authorization thus permits a complex identification and distinction of government and citizen which does fuller justice to this relationship than such metaphors as contract, or Locke's metaphor of trust.[1] Classical liberal theory emphasizes the distinction of government and citizen, conceiving the government as a restraining agent on the citizen. Classical socialist theory emphasizes the identification of government and citizen, conceiving the government strictly as the agent of the body of citizens. I would suggest that the metaphor of authorization brings out both the basic truth which underlies each of these theories, and reveals the limitations of each truth.

Finally, we may assign authorization a useful normative role. Only when the relationships which it requires do actually obtain is government legitimate. Only when the government is effectively the agent of the people, although distinct from them, is obedience to political authority fully obligatory.

Legitimate political power, or political right, comes from the people, in the sense that it must be exercised in their behalf, to secure those of their objectives which cannot be attained by their individual actions. To say this is not, however, to restrict the institutions whereby legitimate political power is concentrated and exercised. It is not to insist on the superiority of representative democracy as a system of government.

[1] Cf. J. W. Gough, *Locke's Political Philosophy*, Oxford, 1950, study vii, for a discussion of the metaphor of trust.

The sovereign must represent the citizens. But this representation need not be formalized in terms of periodic elections, or even periodic plebiscites. As long as the mode of representation is recognized and accepted, and as long as the members of society regard their sovereign as exercising their rights on their behalf, the criteria of legitimate representation are fulfilled. No doubt there are no precise rules for determining whether these criteria are fulfilled. Yet some practical discrimination is surely possible. The governments both of the United States and the Soviet Union are clearly legitimate in terms of these criteria. So are, at the time of writing (May 1967), the governments of the Federal Republic of Germany, Tanzania, and Mexico; so are *not* the governments of the German Democratic Republic, Rhodesia, and Haiti.

V

GOD

To complete our study of Hobbes's moral and political
system we must determine the role played by God. We treat
God last, partly because Hobbes does so, partly because God
plays only a secondary part in the system. We have shown
that neither the formal structure of Hobbes's theory, expressed
in our formal definitions and the conclusions we have drawn
from them, nor the material content, expressed in our material
definitions and their consequences, depends in any important
way on theistic presuppositions. Hobbes's theory is intended
for rational men who aim at their own preservation, whatever
their religious views may be.

The positive interpretation offered in the preceding chapters is our main argument against those recent critics, such
as Taylor, Warrender, and Hood, who consider theism central
to Hobbes's position. But we must show also that what
Hobbes says about God does not require us to revise that
interpretation. For Hobbes insists:

> It is manifest enough, that when a man receiveth two contrary
> commands, and knows that one of them is God's, he ought to obey that,
> and not the other, though it be the command even of his lawful sove-
> reign . . ., or the command of his father. (*E.W.* iii, p. 584)

And so:

> There wants only, for the entire knowledge of civil duty, to know
> what are those laws of God. For without that, a man knows not, when
> he is commanded any thing by the civil power, whether it be contrary
> to the law of God, or not: and so, either by too much civil obedience,
> offends the Divine Majesty; or through fear of offending God, trans-
> gresses the commandments of the commonwealth. (*E.W.* iii, p. 343)

Hobbes never questions the theoretical supremacy of the
authority of God to human authority. This does not mean,
as Warrender would have it, 'that the only power that can

produce obligation is God's power, and never the power of men'.[1] It does mean that in any conflict God takes precedence. What we shall show is that this supremacy does not receive any satisfactory recognition in Hobbes's conceptual scheme, but that it could be accommodated with no basic revision to that scheme, and that it both receives and requires only a very minor alteration in the material content of Hobbes's moral and political system.

It is no part of our position that the traditional interpretation of Hobbes as an atheist should be reinstated against Hobbes's own theistic avowals. In *Human Nature* Hobbes contends:

. . . for the effects we acknowledge naturally, do include a power of their producing, before they were produced; and that power presupposeth something existent that hath such power: and the thing so existing with power to produce, if it were not eternal, must needs have been produced by somewhat before it, and that again by something else before that, till we come to an eternal, that is to say, the first power of all powers, and first cause of all causes: and this is it which all men conceive by the name of GOD, implying eternity, incomprehensibility, and omnipotency. And thus all that will consider, may know *that* God is, though not *what* he is: even a man that is born blind, though it be not possible for him to have any imagination what kind of thing fire is; yet he cannot but know that somewhat there is that men call fire, because it warmeth him. (*E.W.* iv, pp. 59–60)

This argument, repeated in *Leviathan* (cf. *E.W.* iii, pp. 92, 95–6), is surely plain enough. Yet it has been objected that it is impossible to know *that* something is without knowing *what* it is, so that Hobbes in denying all knowledge of what God is, is by consequence an atheist. But although it may be impossible to know that something is without knowing what it is, yet Hobbes did not suppose it impossible. Other arguments have been presented to show that Hobbes's system has various atheistical consequences, but these fail completely to prove that Hobbes is an atheist. He himself implicitly answers all such objections when he says:

When it happeneth that a man signifieth unto two *contradictory* opinions, whereof the *one* is *clearly* and directly *signified*, and the

[1] Howard Warrender, 'A Reply to Mr. Plamenatz', in *Hobbes Studies*, p. 91.

other either *drawn* from that by *consequence*, or not known to be contradictory to it; then, when he is not present to explicate himself better, we are to take the *former* for his opinion; for that is clearly signified to be his, and directly; whereas the other might proceed from error in the deduction, or ignorance of the repugnancy. (*E.W.* iv, pp. 75–6)

Hobbes's theistic views are explicit. His allegedly atheistic views are at best drawn by consequence from doctrines which he does not recognize to contradict his theism. We are therefore to take theism for his opinion.

I. THE NATURAL KINGDOM OF GOD

All of creation is necessarily subject to divine omnipotence. But this power, which operates through the laws discovered by natural science, is to be distinguished from that power which is appropriate to a ruler. Creation does not constitute a kingdom, except metaphorically. 'For he only is properly said to reign, that governs his subjects by his word, and by promise of rewards to those that obey it, and by threatening them with punishment that obey it not' (*E.W.* iii, p. 344).

The kingdom of God comprises only those over whom God rules by his words, and not by mere power. Hobbes distinguishes 'a triple word of God, *rational*, *sensible*, and *prophetic*' (*E.W.* iii, p. 345), given respectively 'by the dictates of *natural reason*, by *revelation*, and by the *voice* of some *man*, to whom by the operation of miracles, he procureth credit with the rest' (ibid.). Thus there are three possible ways in which God might rule by his words.

But to rule God must not only speak, but give laws through his speech. And Hobbes insists, apparently as a matter of fact, that no laws are given in God's sensible word, 'because God speaketh not in that manner but to particular persons, and to divers men divers things' (ibid.). Both God's rational word and his prophetic word are the source of laws, and so he may be said to rule in two, quite distinct ways.

From the difference between the other two kinds of God's word, *rational*, and *prophetic*, there may be attributed to God, a twofold kingdom, *natural*, and *prophetic*: natural, wherein he governeth as many of mankind as acknowledge his providence, by the natural dictates of right reason; and prophetic, wherein having chosen out one peculiar nation, the Jews, for his subjects, he governed them, and none but them, not only by natural reason, but by positive laws, which he gave them by the mouths of his holy prophets. (Ibid.)

God's prophetic kingdom has now lapsed; it will be restored by Christ at the Second Coming, and all who achieve salvation will then be subjects in it.

It is with God's natural kingdom that we are primarily concerned. His prophetic kingdom is a commonwealth, in which he is literally sovereign; hence in it religion and politics are one. There is no human authority in it which might conflict with divine authority. But God's natural kingdom coexists with the rule of human sovereigns. We must then determine the role of God as ruler of his natural kingdom to see whether it is compatible with the political system with which it coexists.

A. *God's Natural Right to Rule*

Hobbes's attempt to establish God's natural sovereign right is found in a long passage which must be considered in full:

The right of nature, whereby God reigneth over man, and punisheth those that break his laws, is to be derived, not from his creating them, as if he required obedience as of gratitude for his benefits; but from his *irresistible power*. I have formerly shown, how the sovereign right ariseth from pact: to show how the same right may arise from nature, requires no more, but to show in what case it is never taken away. Seeing all men by nature had right to all things, they had right every one to reign over all the rest. But because this right could not be obtained by force, it concerned the safety of every one, laying by that right, to set up men, with sovereign authority, by common consent, to rule and defend them: whereas if there had been any man of power irresistible, there had been no reason, why he should not by that power have ruled and defended both himself, and them, according to his own discretion. To those therefore whose power is irresistible, the dominion of all men adhereth naturally by their excellence of power; and consequently it is from that power, that the kingdom over men, and the right of afflicting men at his pleasure, belongeth naturally to God Almighty; not as Creator, and gracious; but as omnipotent. (*E.W.* iii, pp. 345–6)

This passage makes clear the meaning of Hobbes's famous doctrine that '*Power irresistible justifies all actions, really and properly*, in whomsoever it be found; less power does not, and because such power is in God only, he must needs be just in all actions' (*E.W.* iv, p. 250). An action is justified if it is performed with right. But right is wanting only when it has been renounced, and it is renounced only for want of power. Hence, in so far as man's power is sufficient, not just to enable him to perform some action successfully, but to assure him that its consequences are fully within his control, he can have no reason to renounce his right to perform it, and hence his power justifies him should he perform it. Only irresistible power is always sufficient to assure a man that the consequences of his actions are fully within his control, and so only irresistible power always justifies a man in whatever he does.

But this argument is insufficient to show that God possesses sovereign right. In the first place, it will be noticed that Hobbes reverts to his initial theory of sovereignty—the sovereign is not the universal actor, exercising the rights of all of his subjects as authors, but the exclusive possessor of natural right, all other persons having abandoned their rights. This is not, however, the chief weakness in the argument. Sovereign right is not merely permissive; it must be binding on those over whom it is exercised. In this respect it is not merely a special case of natural right. Sovereign right entails an obligation on the part of the subjects to obey, and Hobbes establishes no such right in the passage quoted.

We must distinguish carefully between what Hobbes takes as the basis of God's right to rule, and what must be the basis of sovereign right. God's right to rule is established if every other person lays down his own natural right to *rule*. But sovereign right is established if every other person either lays down his own right to *resist* the sovereign (the initial theory), or if every other person obliges himself to authorize the sovereign. Laying down one's natural right to rule obliges one only not to command others. It does not oblige one to obey, or even not to oppose, anyone who retains the natural right to rule. On the other hand, laying down one's right to resist some person, or obliging oneself to authorize some person, does as we have seen oblige one to obedience to that person.

Thus to show that God's natural right to rule over all mankind is properly sovereign right, Hobbes must show that all men have an obligation to obey God. His attempt to do this we must postpone until later in this chapter, for it raises some of the most vexed questions in Hobbesian interpretation. But one point should be emphasized here. The question of man's obligation to God is never discussed in *Leviathan*. Hence in *Leviathan* Hobbes never successfully shows that God possesses sovereign right in his natural kingdom.

We conclude, then, that Hobbes shows only that God possesses the right of nature to rule over all others, made effective by his omnipotence, and exclusive by the renunciation, on the part of all others, of their right to rule. Of course, we have shown earlier that sufficient power can compel its recognition as sovereign right—this is the foundation of despotic sovereignty. But Hobbes does not offer that argument in establishing God's right to rule, and so no proof is actually given that God possesses sovereign right over all men.

B. *Divine Law*

If God is to rule in his natural kingdom, then not only must he have sovereign right, but he must exercise that right in the making of laws. The mere right to rule, without a corresponding obligation to obey, is not sufficient for the making of laws, since law is a command 'only of him, whose command is addressed to one formerly obliged to obey him' (*E.W.* iii, p. 251). Hence we must for the present reserve our decision whether God's commands in his natural kingdom are properly laws. In establishing only God's right to rule, Hobbes has not provided sufficient grounding for divine law.

Hobbes distinguishes those laws which concern 'the natural duties of one man to another' and those which concern 'the honour naturally due to our Divine Sovereign. The first are the same laws of nature, of which I have spoken already . . .'

(*E.W.* iii, p. 348). Hobbes goes on to consider the second group of divine laws, but we need not follow him. Our concern is with how we can know that the laws of nature are in fact also divine laws.

That this is a problem for Hobbes is recognized even by those critics who wish to insist that the laws of nature, considered as divine laws, provide the necessary foundation of Hobbes's moral and political system. A. E. Taylor admits, 'I do not know . . . how . . . Hobbes supposes persons unacquainted with the Scriptures to have discovered that the natural law *is* a command of God'.[1]

Hobbes seems to suggest three different ways, in which the laws of nature might be known to be divine. We examine each in turn.

1. The laws of nature are known to be divine laws, because God is known to be the author of nature.

This view is supported only in the earliest of Hobbes's major political writings, *De Corpore Politico*.

And forasmuch as law, to speak properly, is a command, and these dictates, as they proceed from nature, are not commands, they are not therefore called laws, in respect of nature, but in respect of the author of nature, God Almighty. (*E.W.* iv, p. 109)

The laws mentioned in the former chapters, as they are called the laws of nature, for that they are the dictates of natural reason, . . .; so are they also divine laws in respect of the author thereof, God Almighty; . . . (*E.W.* iv, p. 111)

But it is rejected in *Leviathan*:

whose [God's] laws, such of them as oblige all mankind, in respect of God, as he is the author of nature, are *natural*; and in respect of the same God, as he is King of kings, are *laws*. (*E.W.* iii, pp. 342–3)

To recognize God as the author of nature is, it seems, simply to ascribe an ultimate divine origin to all things. Thus such recognition in no way affects the status to be assigned to the laws of nature. If, as Hobbes admits in the first passage we quoted, 'as they proceed from nature' they 'are not commands',

[1] A. E. Taylor, op. cit., p. 50.

then recognizing God as the author of nature will not turn the
laws of nature into divine commands. We have maintained
that the laws of nature are rational precepts; we may then
term them divine precepts instead of natural precepts, but
precepts they remain. What is wanted is an argument to show
that God *commands* adherence to these precepts, but this the
appeal to God as author of nature does not do.

It is possible to interpret recognition of God as the author
of nature somewhat differently, although it does not affect
our argument. One might suppose that to recognize God as
author of nature is to recognize nature as a system regulated
by objective causal principles.[1] Then this would affect the
status to be assigned the laws of nature. Initially, they would
be regarded as no more than rule-of-thumb guides to success
in survival. But once they were seen to be based on an objective causal order in nature, they would be regarded as necessary
precepts, essential to success in survival.

The laws of nature might then prove compelling in either
of two ways. They might be psychologically compelling, so
that a man who recognized that they prescribed the means
necessary to survival in a fixed natural order, would find
himself unable to act contrary to them. They might then be
regarded as laws, but as descriptive laws of behaviour, not as
prescriptive laws commanded by some authority.

Or the laws of nature might be rationally compelling. They
might then be regarded as obliging, in Hobbes's rare sense of
rational obligation, but not as obligatory, nor as laws, since
they would not be the commands of some authority. Thus this
reinterpretation of the recognition of God as author of nature
affords no grounds for supposing that God commands adherence to the precepts which constitute the law of nature, and
so it fails as a defence of Hobbes's first way of showing the
laws of nature to be divine laws.

2. The laws of nature are known to be divine laws, because
they come from reason, which God has given man as a guide.

[1] This view is suggested to me by passages in M. M. Goldsmith, *Hobbes's
Science of Politics*, New York, 1966, especially pp. 132–3.

One can find support for this view in all three of Hobbes's major political writings:

> Finally, there is no law of natural reason, that can be against the law divine: for God Almighty hath given reason to man to be a light unto him. (*E.W.* iv, p. 116)
>
> Because the *word of God*, ruling by nature only, is supposed to be nothing else but right reason, . . . (*E.W.* ii, p. 209)
>
> . . . he governeth as many of mankind as acknowledge his providence, by the natural dictates of right reason; . . . (*E.W.* iii, p. 345)

Wernham seems to combine this view with the first one: 'God then is the author of the law of nature in this sense: that by determining the nature of man he has also determined certain conditions which are necessary to men's living together in peace and lasting security, and by giving man reason he has enabled him to discover these conditions and make them into rules for his conduct.'[1]

But the same objection applies as before. If the laws of nature are not commands as they proceed from nature, then natural reason discerns them as precepts and not commands. To introduce God as giving reason to man as a guide, is then to confirm the status of the natural laws as precepts. It provides no ground whatsoever for supposing that, because God has given man reason to guide him, God must be supposed to command as laws what reason ascertains as precepts.

There is a further objection. To claim that God has given man reason as a guide is to presuppose Divine providence—as indeed the quotation from *Leviathan* suggests. But Divine providence cannot be demonstrated by natural reason. Hobbes insists that 'there is no natural knowledge of man's estate after death' (*E.W.* iii, p. 135). Hence the second way cannot suffice for establishing that the laws of nature are divine precepts, much less divine laws.

Unfortunately for Hobbes's account of God's natural kingdom, and for those interpreters who insist that the laws of nature as commands of God are essential to Hobbes's moral and political theory, the 'rational word of God' cannot

---

[1] A. G. Wernham, op. cit., p. 136.

be known, by reason alone, to be the word of God, and nothing delivered in it has the character of a law.[1] What is more, Hobbes admits this in his *Answer to Bramhall*: 'I thought it fittest in the last place, once for all, to say they [the laws of nature] were the laws of God, then when they were delivered in the 'word of God; but before, being not known by men for any thing but their own natural reason, they were but theorems, tending to peace, . . .' (*E.W.* iv, pp. 284–5) and in *Behemoth*: '. . . men can never by their own wisdom come to the knowledge of what God hath spoken and commanded to be observed, . . .' (*E.W.* vi, p. 221).

If natural reason is distinguished, as in the *Answer to Bramhall*, from the word of God, then there is no rational word of God. If natural reason is precluded, as in *Behemoth*, from discovering the commands of God, then there is no natural kingdom of God. Hobbes can escape from these conclusions only by inconsistency.

3. The laws of nature are known to be divine laws, because they are found in Scripture.

This view admits the objections we have raised to the possibility of natural knowledge of divine law. It is supported only in *De Cive*: 'But those which we call the laws of nature, . . . are not . . . laws, as they proceed from nature. Yet, as they are delivered by God in holy Scriptures, . . . they are most properly called by the name of laws. For the sacred Scripture is the speech of God commanding over all things by greatest right' (*E.W.* ii, pp. 49–50).

If the laws of nature can be shown to be divine laws only by an appeal to Scripture, then they cannot be the laws of God's natural kingdom. They are laws only for Christians, and in Hobbes's view acceptance of Christianity depends not on reason but on revelation, faith, and authority.

We may, however, now introduce a new kingdom which we shall call God's scriptural kingdom. It will differ from God's

---

[1] Heroically, of course, one can draw the conclusion that Hobbes's system presupposes Christianity, as F. C. Hood does. If our interpretation is correct, Christianity is important for Hobbes only in so far as it must be reconciled with his views, which are themselves entirely independent of it.

natural kingdom in its basis, since it depends on God's prophetic word as found in Scripture, and not on his natural word. It will differ also in including Scriptural laws for the worship of God, and not merely the dictates of natural reason for divine honour. But its other laws will be just the Hobbesian laws of nature, given that as Hobbes insists, these may be found in Scripture.

The great majority of Hobbes's readers, being Christians, will find it a matter of basic indifference whether they suppose themselves to be members of God's scriptural kingdom or God's natural kingdom. In either case, God's sovereignty will require reconciliation with human sovereignty, and God's sovereign right will have to be fitted into Hobbes's essentially secular moral and political system. Thus Hobbes's arguments about God will have much the same significance whether we suppose the discussion to be about God's natural kingdom or our proposed scriptural kingdom.

2. GOD'S SUBJECTS

Who are to be included among God's subjects? Evidently 'not bodies inanimate, nor creatures irrational' (*E.W.* iii, p. 344), for they cannot be ruled by God's word. They can be subject only to the laws of physical nature, not to the laws of political community. Thus only men, and men of sound mind, can be members.   '

But not all men of sound mind are among God's subjects. Atheists are not, for not believing that there is a God they cannot recognize any rules as his commands. Atheists are God's enemies.

Atheism Hobbes considers a sin, but a sin '*of imprudence or ignorance*' (*E.W.* ii, p. 198). Although natural reason is sufficient to prove the existence of God, yet not all men are capable of knowing this—just as, Hobbes points out (ibid.), not all men can know the proportion of the circle to the square, although Archimedes discovered it by natural reason. But ignorance does not excuse the atheist. For in the state of nature any man may rightly kill any other; the atheist, not recognizing God's rule, is in the state of nature with respect to him, and so may rightly be killed by him as an enemy.

Men who, though not atheists, 'believe not that God has any care of the actions of mankind' (*E.W.* iii, p. 344), are also not among God's subjects. To deny God's providence, as much as to deny God's existence, leaves one in the state of nature with respect to God, and so makes one his enemy.

Since as we have seen, Divine providence is not knowable by natural reason, this once again reveals the incoherence in Hobbes's account of God's natural kingdom. Men who believe in providence, whether they know it or not, will of course be God's subjects, but this belief will not be guaranteed by natural reason. If membership in God's natural kingdom were a condition of possessing rights and obligations, or of membership in political society, this would be a fatal weakness in Hobbes's theory.

All rational theists who believe in divine providence are to be taken as God's subjects. We must now inquire into the basis of their obligation to obey God. As we have already noted, this is a subject considered only in *De Cive*. Since *De Corpore Politico* contains no mention of God's natural kingdom, its omission there is hardly surprising. What is surprising is its omission in *Leviathan*.

This omission is almost certainly deliberate. *De Cive* and *Leviathan* are not markedly similar works; there is no reason to suppose that, when Hobbes came to write *Leviathan*, he had *De Cive* to hand. But chapter 31 of *Leviathan*, which treats of the natural kingdom of God, is precisely parallel to the corresponding chapter xv of *De Cive*, from the beginning to the point at which man's obligation to obey God is introduced in *De Cive*. There is, I think, no reasonable possibility that this part of *Leviathan* could have been composed without direct and detailed reference to *De Cive*. And the reader may easily satisfy himself on this point by making his own careful comparison of the two chapters.

Furthermore, the parallelism between *Leviathan* and *De*

Cive resumes, although only briefly, immediately after the discussion of man's obligation to obey God. But *Leviathan* simply omits that discussion itself. I am, therefore, satisfied that Hobbes, when he came to the discussion of that obligation in *De Cive*, must have chosen to omit it from *Leviathan*.

We cannot yet attempt to answer the question why Hobbes chose to omit it. Once we have examined the discussion, we shall see that Hobbes would have had good reason to consider it entirely unsatisfactory, without having any alternative account easily forthcoming. For the present we need note only that the discussion in *De Cive*, which we shall now quote, cannot be relied on as a statement of Hobbes's final view of man's obligation to obey God—there is no such statement, and there may well be no 'final view':[1]

> Now if God have the right of sovereignty from his power, it is manifest that the *obligation* of yielding him obedience lies on men by reason of their weakness. For that *obligation* which rises from contract, of which we have spoken in chap. II. can have no place here; where the right of ruling, no covenant passing between, rises only from nature. But there are two species of *natural obligation*. One, when liberty is taken away by corporal impediments, according to which we say that heaven and earth, and all creatures, do obey the common law of their creation. The other, when it is taken away by hope or fear, according to which the weaker, despairing of his own power to resist, cannot but yield to the stronger. From this last kind of obligation, that is to say, from fear or conscience of our own weakness in respect of the divine power, it comes to pass that we are obliged to obey God in his natural kingdom; reason dictating to all, acknowledging the divine power and providence, *that there is no kicking against the pricks*. (*E.W.* ii, p. 209)[2]

The first sense of natural obligation is a red herring, in no way germane to Hobbes's argument. For although we may use the word 'oblige' here, and say, for example, that if a stone is dropped it is *obliged* to fall to the earth, we cannot say that the

---

[1] I find it astonishing that Hobbes's commentators should not have considered the significance of the omission of any account of man's obligation to obey God from *Leviathan*. It is surely a major stumbling block for anyone who supposes such an obligation to be central to Hobbes's system.

[2] My discussion of this passage owes much to A. G. Wernham, op. cit., pp. 130–2, and John Plamenatz, op. cit., pp. 83–5.

stone has an *obligation* to fall to the earth, nor that it wants the *right* or *liberty* to soar up into the skies. The first sense of natural obligation concerns only want of capacity or power, not want of liberty or right. Hobbes makes this clear when he says: 'And though the water cannot ascend, yet men never say it wants the *liberty* to ascend, but the *faculty* or *power*, because the impediment is in the nature of the water, and intrinsical' (*E.W.* iv, p. 274).

What of the second sense of natural obligation, the one relevant since it is the sense in which Hobbes insists we are obliged to obey God? It is by no means clear how it is to be understood, and I want to suggest three possible meanings which depend on three possible interpretations of the crucial phrase 'cannot but yield'.

1. We may take 'cannot but yield' to indicate *psychological compulsion*. On this view, the weaker, aware of his weakness, is psychologically unable to do anything but obey the stronger. This is, I think, the interpretation Warrender places on the obligation to obey God, for he says:

> Now if irresistible power is related to moral obligation it will be irresistible political power, and may be described as something in the patient or subject that *necessarily* determines his will on reflection. The thought of God or of his sanctions necessarily moves those who deliberate adequately to act in conformity with what they take to be God's will.[1]

Now if this is the correct interpretation of 'cannot but yield', the second sense of natural obligation reduces to a special case of the first sense. The first sense covers all cases in which some object acts, as it must, in accordance with physical law; the second sense covers that particular case in which some man acts, as he must, in accordance with a particular physical law, in this case a law of Hobbesian physiological psychology. It follows that on this interpretation a man may be said to be obliged to obey God, because he *can* do no other. But he will not have any obligation to obey God because no question of what he *may* do, or has the *right* to do, arises.

[1] Howard Warrender, *The Political Philosophy of Hobbes*, p. 315.

This account of the obligation to obey God correlates it with that view of the laws of nature which takes them to be descriptive laws of behaviour, in so far as they are considered as part of the divinely created order of nature. It has nothing whatsoever to do with any sort of *practical* obligation, moral or prudential. That man has an obligation to obey God is not then a prescriptive conclusion of Hobbes's moral and political theory, but a descriptive conclusion from his psychology, relevant to morals and politics only in that it sets out a psychological limitation on possible human action.

2. We may take 'cannot but yield' to indicate *rational necessitation*. On this view, the weaker, aware of his weakness, is rationally unable to do anything but obey the stronger. If he fails to obey, then he is irrational. It may be that Warrender intends to place this interpretation on the obligation to obey God, although I do not find it easy to read his words in this way.

Now if this is the correct interpretation of 'cannot but yield', the second sense of natural obligation reduces to a species of rational obligation. As we have seen before, rational obligation is not a restriction of right; rather it is a determination of right. For Hobbes, a right to do what is contrary to reason would be impossible by definition; rational obligation determines what must be done to accord with reason. It follows that on this interpretation a man may be said to be obliged to obey God, but not to have a moral (or practical) obligation to obey God. What is rationally obliging is not *obligatory*.

We use 'oblige' in much this sense when we say that a man is *obliged* to give up his money to an armed robber, to avoid being shot.[1] He has no *obligation* to give up his money to the robber, and he *can* do otherwise, but it would be *irrational* for him to do otherwise.

This account of the obligation to obey God suggests a correlation with that view of the laws of nature, which takes

---

[1] The example in this context is adapted from A. G. Wernham, op. cit., p. 131.

them to be rationally compelling precepts of behaviour, in so far as they are considered as part of the divine order of nature. But it does not quite correlate with that view, for the rational compulsion may derive from beliefs about one's fate hereafter, and not merely from beliefs about the natural conditions of preservation. It does have a certain practical relevance, in so far as the person who considers himself obliged to obey God because of fears extending even beyond fear of earthly death, has an additional reason for acting on the laws of nature. And of course, if the obligation to obey God were taken to be only the rational requirement of conforming to the conditions of earthly preservation determined by an objective causal order, then its practical relevance would be assured but at the price of making both 'obligation' and 'God' unnecessary to its formulation.

On this second interpretation, then, obligation to God is to be distinguished from obligation in Hobbes's usual sense, as no more than rational obligation. In so far as reference to God is taken to be essential, it plays no fundamental role in Hobbes's moral and political system. And it clearly cannot serve as the basis for deriving those moral or practical obligations which arise from renunciation of right.

3. We may take 'cannot but yield' to indicate the necessity, either psychological or rational, of *authorizing God* or of *giving up one's right to oppose God*. On this view, the weaker, aware of his weakness, must accept the sovereignty of the stronger. This is the interpretation which Goldsmith seems to accept: 'Man's natural obligation to God, as distinguished from any covenanted obligation, is derived from an act of man's mind, his submission when he understands and acknowledges God's power and providence.'[1]

The difficulty which this interpretation presents is that it does *not* distinguish the obligation to obey God from any covenanted obligation. The obligation has exactly the same basis as the obligation of the vanquished to obey the victor. Plamenatz sees this point clearly: 'For men submit to God

[1] M. M. Goldsmith, op. cit., p. 112.

because of his power, his threats, and his rewards. Their submission to him is voluntary in exactly the same sense as their submission to a conqueror, and Hobbes does not pretend otherwise. In both cases, they choose to submit; . . .'[1]

On this interpretation, then, Hobbes is confronted with a dilemma. Political obligation must rest on covenant. Natural obligation to God must be independent of covenant, for as we have seen: 'To make covenant with God, is impossible, but by mediation of such as God speaketh to, either by revelation supernatural, or by his lieutenants that govern under him, and in his name: . . .' (*E.W.* iii, p. 125). But natural obligation to God is logically equivalent to one type of political obligation. To quote Plamenatz once more:

The obedience or submission, when it proceeds from a just estimate of the situation, is in both cases exactly the same; and if there is a covenant that creates a duty of obedience in the one case, there must also be a covenant in the other. Now, this is a conclusion that Hobbes would never have accepted, for he was as certain that our duty to obey God does not rest on covenant as that our duty to obey man does.[2]

---

[1] John Plamenatz, op. cit., p. 84.

[2] Ibid. But Plamenatz goes on to argue that despotic sovereignty *cannot* be based on covenant. 'We obey, according to Hobbes, only when we know that someone commands us and is powerful. Therefore, all obedience, properly so called, is submission. If, then, submission involves a tacit covenant, so does obedience. In other words, whenever we obey, we promise to obey, and this promise makes obedience a duty. Which is absurd' (p. 85).

But this argument is quite mistaken. Whenever we obey, either we have submitted previously—obliging ourselves to obey—or not. If we have submitted previously, then our obedience is simply a continuation of that submission. No new promise to obey is involved.

Suppose then that we have not submitted previously. We find ourselves confronted with someone powerful, who issues a command to us, presumably leaving us sufficient corporal liberty to obey or not. In such a situation fear leads us to submit, and the sign of our submission will be obedience to the command. This obedience, then, will constitute the offer of a covenant of obedience, an offer which may well be supplemented by other signs of the will. If the offer is accepted—if, that is, we are allowed further corporal liberty—then we are obliged to further obedience. Our obedience is a promise to further obedience, which, if accepted, makes such further obedience obligatory. Which is not absurd at all.

The initial act of obedience is not obligatory, although we may be compelled (obliged) to perform it through fear. Plamenatz constructs his alleged absurdity by confusing the obedience which may constitute our initial submission with the obedience which follows on submission. When they are distinguished, Hobbes's doctrine that submission involves covenant is upheld.

This account of the obligation to obey God correlates it with that view of the laws of nature which takes them to be actual divine commands. It is required by Hobbes to explain the manner in which the laws of nature oblige in conscience. But as we have seen, the effectiveness of the laws of nature in Hobbes's system does not depend on our being obliged in conscience to obey them. It is also required by Hobbes to establish the subordination of the sovereign to God. But this again is not an essential part of Hobbes's system; the role of the sovereign does not depend on his being subordinated to God. Hence the obligation to obey God is theoretically dispensable, although it is certainly false that Hobbes would have happily dispensed with it.

If our analysis of the second sense of natural obligation is correct, we can see why Hobbes found it unsatisfactory, as an account of man's obligation to obey God, and yet found no alternative account. For either the obligation to obey God turns out not to be an obligation at all (first and second interpretations) or it turns out to be indistinguishable from the covenanted obligation to obey the victor (third interpretation). And no plausible alternative account presents itself, because Hobbes's conceptual system has no place for an obligation which is properly a restriction of right, yet is not imposed by covenant or some essentially equivalent device.

Recent interpretations of Hobbes can be understood as attempts to resolve the genuine difficulties which Hobbes faced in constructing his moral and political theory. In particular, the Taylor–Warrender thesis seeks to show how it is possible for Hobbes to claim that we have an obligation to obey God, in the same sense, though not on the same grounds, that we have an obligation to adhere to our covenants, and to obey our earthly sovereign. The Taylor–Warrender thesis does this by reducing all obligation in Hobbes to an ultimate obligation to obey God, from which all else must be derived. Thus it pays the price of denying the secular character of Hobbes's theory.

Unfortunately, the thesis accomplishes nothing, despite the

price it pays. For the reduction works in the wrong way. Instead of showing that we have an obligation to obey God in the same sense as our obligation to keep covenants, the Taylor–Warrender thesis must conclude that we have an obligation to keep covenants in the same sense as our obligation to obey God. The obligation to obey God is then interpreted as psychological and/or rational necessitation: 'to be morally obliged is to be necessarily moved on reflection.'[1] We have the formal definition:

'obligation' = 'necessitation on reflection'

from which unfortunately it follows that:

'obligation' (in Hobbes) ≠ 'obligation' (in ordinary English)

It is not just that Hobbesian obligation is not *moral* obligation; it is not obligation *at all*. To be necessitated on reflection may be to be obliged, but it is not to have an obligation.

Our interpretation of Hobbes avoids this fatal conclusion. We have shown that Hobbes's concept of obligation is recognizably similar to our own. But we are left with the problem, albeit a minor problem, that Hobbes seems to have no satisfactory account of man's obligation to obey God.

We shall therefore construct such an account. To do so, we shall sacrifice Hobbes's insistence that the obligation to obey God does not rest on covenant. But we shall show that this sacrifice involves no theoretical departure from Hobbes's system.

Let us first ask why Hobbes is committed to the view that man's obligation to obey God in his natural kingdom does not rest on covenant. Hobbes gives no specific reasons, but he

---

[1] Howard Warrender, op. cit., p. 315. The passage quoted is preceded by an 'if'; Warrender would prefer Hobbes to base obligation 'upon a body of natural law having self-evident or intrinsic authority', thus making 'it unnecessary to introduce the role of God at all into Hobbes's political theory' (*Hobbes Studies*, p. 90). This would have the merit of restoring the secular character of Hobbes's theory, at the cost of turning the theory into one both totally different from and much less plausible than Hobbes's actual theory.

does present two general objections to covenant between man and God, except by mediation of God's representative. The first is that without such mediation 'we know not whether our covenants be accepted, or not' (*E.W.* iii, pp. 125–6). The second is that such covenants would provide a pretence for disobedience to the sovereign:

And whereas some men have pretended for their disobedience to their sovereign, a new covenant, made, not with men, but with God; this also is unjust: for there is no covenant with God, but by mediation of somebody that representeth God's person; which none doth but God's lieutenant, who hath the sovereignty under God. (*E.W.* iii, pp. 160–1)

Neither of these two objections really applies to the proposal that man covenants to authorize God in his natural kingdom. The second objection may be disposed of immediately. Hobbes supposes that man has an obligation to obey God; to rest that obligation on covenant would alter the ground, but not the content, of the obligation. Hence if a man can pretend a right to disobey his earthly sovereign, because of his obligation to obey God in his natural kingdom, this pretence is quite independent of whether the obligation to God is grounded on covenant. We shall consider in the next part of this chapter whether God's natural kingdom can conflict in any way with earthly kingdoms, but the possibility of conflict, if it exists, has no bearing on our present argument.

The first objection is met by a consideration of what would be involved in the acceptance by God of man's covenant of authorization or obedience in his natural kingdom. The parallel is with acceptance by the despotic sovereign of his subjects' covenants of authorization, and here we must recall the technically degenerate nature of these covenants. The sovereign transfers no rights; the obligation to keep the covenants turns solely on whether he allows his subjects life and corporal liberty. Thus his acceptance is signified if, and as long as, he allows them this life and liberty.

Similarly, God's acceptance of a covenant of obedience in his natural kingdom would be signified by his allowing men life and corporal liberty. There is a complication, in that through obedience to God man seeks not only earthly life, but salvation or eternal life, as opposed to damnation which Hobbes construes as eternal death (cf. *Leviathan*, chapter 38). No man can know in this present life whether he will be saved or damned. But every man can hope to be saved, and so every man must act as if his covenant of obedience to God is accepted. Thus God's acceptance would be signified by his not actually damning man, and this raises no problems.

Thus Hobbes has no adequate grounds, in terms of his theory, for denying that man's obligation to obey God in his natural kingdom rests on man's covenant authorizing God as his natural sovereign. And so we suggest that such a covenant might be incorporated into Hobbes's theory, and that thereby man's obligation to obey God would be put on the same basis as his obligation to keep covenants and to obey his earthly sovereign.

Such a covenant would have two peculiar features. Unlike other covenants of submission it would not preclude the making of a further covenant among men to submit to an earthly sovereign. For this covenant would not institute commonwealth, but only the natural kingdom of God. The natural kingdom of God does not provide its subjects with security against each other, or against other men, but only the hope of security against God. And thus it does not by itself make men secure; only civil society can do this.

Second, the covenant authorizing God's natural sovereignty unlike all other covenants, could not be invalidated by the civil sovereign. But this is no objection, for God's natural kingdom cannot be denied by the civil sovereign. Men are God's subjects, in Hobbes's view, whether the sovereign will it or not. Again, we shall consider whether this gives rise to any conflict between God's natural kingdom and civil society, in the next part of this chapter.

The only conclusions we can draw from Hobbes's account of man's obligation to obey God is that it is not satisfactory as it stands, that Hobbes seems to be aware that it is not satisfactory, but that a satisfactory account could be provided at no real cost to Hobbes's moral and political system. However, since Hobbes does not require God, and hence any obligation to obey God, in order to establish his moral and political doctrines, we need not find our rather negative conclusions unduly disconcerting.

3. RELIGION AND COMMONWEALTH

A. *The Natural Kingdom of God and Commonwealth*

God's natural sovereignty extends over all who accept his power and providence—sovereigns and subjects, men in civil society and men in the state of nature. But his yoke is light, for in the state of nature he merely reinforces those precepts which set out the means necessary for preservation, and in civil society he merely reinforces the authority of the civil sovereign.

The civil sovereign is God's representative, as far as his subjects are concerned. In all matters relating to religion, the law of the commonwealth is authoritative:

There is no nation in the world, whose religion is not established, and receives not its authority from the laws of that nation. It is true, that the law of God receives no evidence from the laws of men. But because men can never by their own wisdom come to the knowledge of what God hath spoken and commanded to be observed, nor be obliged to obey the laws whose author they know not, they are to acquiesce in some human authority or other. So that the question will be, whether a man ought in matter of religion . . . to rely upon the preaching of his fellow-subjects or of a stranger, or upon the voice of the law? (*E.W.* vi, p. 221)

We shall see in the next section that this overstates Hobbes's position in one important respect. But as far as natural religion, and the natural kingdom of God, are concerned, Hobbes admits no exceptions to his doctrine that the civil sovereign is to his subjects the authority to be followed.

Indeed, Hobbes goes so far as to insist that the sovereign is the public conscience, determining for his subjects what is good and evil in the sight of God. Whatever is done against conscience is sin, but

> ... the conscience being nothing else but a man's settled judgment and opinion, when he hath once transferred his right of judging to another, that which shall be commanded, is no less his judgment, than the judgment of that other. So that in obedience to laws, a man doth still according to his own conscience, but not his private conscience. (*E.W.* iv, pp. 186–7)

And this leads to the conclusion:

> Therefore, though he that is subject to no civil law, sinneth in all he does against his conscience, because he has no other rule to follow but his own reason; yet it is not so with him that lives in a commonwealth; because the law is the public conscience, by which he hath already undertaken to be guided. (*E.W.* iii, p. 311)

There is no possibility that a man's loyalty might be divided between God's natural kingdom and his civil commonwealth. Within the framework of civil society the role of God's natural kingdom is to add a religious character to civil obligations and duties. This extends to the sovereign as well as the subject—both are bound more effectively to their proper roles by their belief that they are obliged to obey God in his natural kingdom.

But it may be asked, what if the sovereign refuses to enforce the laws of nature which are also the divine laws? Hobbes refuses to admit such a possibility.

> ... the laws of nature, ... in the condition of mere nature, ... are ... but qualities that dispose men to peace and obedience. When a commonwealth is once settled, then are they actually laws ...; as being then the commands of the commonwealth; and therefore also civil laws: for it is the sovereign power that obliges men to obey them. ... The law of nature therefore is a part of the civil law in all commonwealths of the world. (*E.W.* iii, p. 253)

If we accept Hobbes's claim that the laws of nature are confirmed by Scripture, then Hobbes's political system is compatible not only with the natural kingdom of God but also with what we have proposed as the scriptural kingdom of God. But Scripture contains further tenets, relating not to man's preservation on earth but to his eternal salvation. These tenets go beyond natural reason, but they are part of the religion of most of his readers, and so Hobbes is careful to show also their compatibility with his doctrines. Although it is only peripheral to our inquiry, we shall therefore consider in the next section Hobbes's arguments on this subject. But anyone whose concern is solely with the positive structure and content of the moral and political system may safely omit this section.

B. *Salvation and Commonwealth*

The question whether the conditions laid down in Scripture for salvation are compatible with obedience to the civil sovereign may be considered equally as the question whether God's prophetic kingdom is compatible with commonwealth. For salvation, or eternal life, is equated by Hobbes with reception into the prophetic kingdom of God in which Christ shall rule at the Second Coming. Hobbes argues:

> But this difficulty of obeying both God and the civil sovereign on earth, to those that can distinguish between what is *necessary*, and what is not *necessary for their reception into the kingdom of God*, is of no moment. For if the command of the civil sovereign be such, as that it may be obeyed without the forfeiture of life eternal; not to obey it is unjust; . . . But if the command be such as cannot be obeyed, without being damned to eternal death; then it were madness to obey it, . . . (*E.W.* iii, p. 585)

Hobbes's account of what is necessary to salvation is engagingly simple: 'All that is NECESSARY *to salvation*, is contained in two virtues, *faith in Christ*, and *obedience to laws*' (ibid.). Obedience raises no problems, for Hobbes finds that 'our Saviour Christ hath not given us new laws, but counsel to observe those we are subject to; that is to say, the laws of nature, and the laws of our several sovereigns' (*E.W.* iii, pp. 586–7). Precepts of Scripture which are not part of the law of nature are law at the discretion of the civil sovereign, and otherwise counsel; hence no conflict can arise between them and the laws of commonwealth.

What then is required for faith in Christ? 'The *unum neces-sarium*, only article of faith, which the Scripture maketh simply necessary to salvation, is this, that JESUS IS THE CHRIST' (*E.W.* iii, p. 590). Hobbes endeavours to prove this, showing at some length that all other belief is either a mere corollary of this, or not necessary to salvation.

Any Christian sovereign will, and must, allow and indeed propagate this belief. Therefore obedience to a Christian sovereign can never conflict with what is necessary to salvation. Even should the sovereign err in the consequences he draws from the basic article of faith, yet private men are not to judge of such error, and are not in danger of damnation if they acquiesce in it.

An infidel sovereign may not allow the public acceptance or proclamation of Christian belief. But this is fortunately no matter. Obedience to the laws of commonwealth is commanded by Christ, whether the sovereign be Christian or infidel, and faith in Christ needs no public acceptance or proclamation, for 'it is internal and invisible' (*E.W.* iii, p. 601). It is not a man's words, nor his deeds, but his thoughts, which determine whether he has the faith necessary to salvation, and thoughts cannot be subject to the control of the sovereign (cf. *E.W.* iii, p. 493).

But what if the infidel sovereign demands that his subjects publicly deny faith in Christ? Then, Hobbes says, 'they have the license that Naaman had' (*E.W.* iii, p. 601).

. . . Naaman believed in his heart; but by bowing before the idol Rimmon, he denied the true God in effect, as much as if he had done it with his lips. But then what shall we answer to our Saviour's saying, (*Matth.* x. 33) *Whosoever denieth me before men, I will deny him before my Father which is in heaven.* This we may say, that whatsoever a subject, as Naaman was, is compelled to do in obedience to his sovereign, and doth it not in order to his own mind, but in order to the laws of his country, that action is not his, but his sovereign's; nor is it he that in this case denieth Christ before men, but his governor, and the law of his country. (*E.W.* iii, pp. 493-4)

Has not Hobbes forgotten that the subject is author of all the sovereign does, and so if the infidel sovereign denies

Christ, the subjects all thereby deny him? This poses a tricky problem. If the subjects do authorize the sovereign denial of Christ, then they have forfeited salvation. If they do not, then in commanding them to deny Christ the sovereign is exceeding his authorization, and so if they obey, they cannot plead an obligation to obey as their excuse, and so they have forfeited salvation. Although Hobbes does not consider this question, it seems that his solution would be that the subjects do not authorize the denial of Christ, but that they have the right to acquiesce in his denial, if the sovereign would punish them for not doing so, since acquiescence is not necessarily damning, and man is not obliged by God to risk punishment or death for doing what is not necessarily damning. As long as the subjects believe in their hearts, they may be saved.

Hobbes qualifies this extreme position in two ways. First, he suggests that an infidel sovereign will not require his Christian subjects to renounce their faith, for

. . . what infidel king is so unreasonable, as knowing he has a subject, that waiteth for the second coming of Christ, after the present world shall be burnt, and intendeth then to obey him, (which is the intent of believing that Jesus is the Christ,) and in the mean time thinketh himself bound to obey the laws of that infidel king, (which all Christians are obliged in conscience to do), to put to death or to persecute such a subject? (*E.W.* iii, p. 602)

Hobbes does not usually display such naïvete. His contemporaries could have enlightened him on this matter, for they recognized implicitly, what we recognize explicitly, that religion is a matter of emotion, as well as, and sometimes instead of, reason.

Second, Hobbes concedes that those persons who have 'a warrant to preach Christ come in the flesh' and 'are sent to the conversion of infidels' (*E.W.* iii, p. 496) must refuse to deny Christ publicly, even if commanded to do so by their sovereign. For their denial would prevent the conversion, and hence salvation, of other men. But they are not to resist the civil sovereign. Rather they must '(g)o to Christ by martyrdom;

which if it seem to any man to be a hard saying, most certain it is that he believes not with his whole heart, *that Jesus is the Christ*' (*E.W.* ii, p. 316). Thus in this one case there is the possibility of conflict between what is required by God for salvation, and what is required by the civil sovereign. The claims of God must then take priority.

What is of interest to us in this discussion of the conditions of salvation is the tacit assumption that the subject does not authorize the sovereign to judge what is requisite to his individual salvation. The subject must acquiesce in the sovereign's interpretation of what God requires of man in his natural kingdom, but not in the sovereign's interpretation of what God requires of man for entry into his prophetic kingdom. It seems then to be only a fortunate fact that what the Scriptures enjoin is compatible with the laws of commonwealth. And so it seems a fortunate fact that religion proves compatible with Hobbes's secular moral and political system.

C. *Conclusion*

At the beginning of this chapter we quoted Hobbes's admission that the commands of God are to be obeyed rather than the commands of man, should the two conflict. We have shown that for Hobbes the two do not conflict. But this is surely a convenient, but not a necessary, fact. It is logically possible that divine authority should conflict with human authority, and Hobbes admits the supremacy of divine authority. But then it may be objected that Hobbes's theory does not have the secular character we have ascribed to it. Although theism may for practical purposes be superfluous, logically it is central to Hobbes's system.

This objection would rest on confusion. In insisting on the secular character of Hobbes's theory, we are making two claims.

1. The formal structure of Hobbes's moral and political theory is independent of any theistic suppositions. This we have shown by establishing formal definitions of Hobbes's moral and political concepts which contain no theistic terms.

2. The material content of Hobbes's moral and political theory is independent of any theistic suppositions. This we have shown by establishing Hobbes's moral and political conclusions using only his formal concepts and his account of human nature, which is devoid of theistic suppositions.

But nothing in Hobbes's theory prevents the introduction of theistic suppositions. If these are introduced, then there are two possibilities.

(a) The conclusions drawn from these suppositions, together with the formal concepts and the account of human nature, are not fully compatible with the conclusions drawn excluding the theistic presuppositions. In this case the material content of the theory would demand revision, and (2), though not (1), would fail.

(b) The conclusions drawn from these suppositions, together with the formal concepts and the account of human nature, are the same as the conclusions drawn excluding the theistic presuppositions. In this case the theistic suppositions are logically superfluous, and (2) as well as (1) holds.

Since Hobbes accepts possibility (b), our insistence on the secular character of his theory is upheld.

We may ask what would be required for Hobbes to accept possibility (a). The conclusions Hobbes draws depend on what is necessary for human preservation. When theistic suppositions are introduced, the conclusions become dependent both on what is necessary for human preservation and for human salvation. Hence possibility (a) would arise if what was required for salvation differed from what was required for preservation.

Now I should argue that Hobbes is prepared simply to refuse to admit the truth of any religious view which would make the conditions for salvation incompatible with the conditions for preservation. We know what promotes life and well-being in this world. We do not know what promotes life and well-being in the hereafter. Hence we shall do well to assume that whatever promotes salvation will not conflict with whatever promotes that temporal end which we naturally and necessarily seek. Any religious view which maintains the contrary we dismiss as superstition.

Hobbes's account of natural religion, and his interpretation of Christianity, could then be defended by showing that they afford support to his secular moral and political system. He would then be in agreement with 'the first founders, and legislators of commonwealths among the Gentiles, whose ends were only to keep the people in obedience, and peace', for 'they have had a care, to make it believed, that the same things were displeasing to the gods, which were forbidden by the laws' (*E.W.* iii, p. 103).

'And thus you see how the religion of the Gentiles was a part of their policy' (*E.W.* iii, p. 105). Hobbes is best understood as a Gentile.

AN APPENDIX

HOBBES ON INTERNATIONAL RELATIONS

But though there had never been any time, wherein particular men were in a condition of war one against another; yet in all times, kings, and persons of sovereign authority, because of their independency, are in continual jealousies, and in the state and posture of gladiators; having their weapons pointing, and their eyes fixed on one another; that is, their forts, garrisons, and guns upon the frontiers of their kingdoms; and continual spies upon their neighbours; which is a posture of war. (*E.W.* iii, p. 115)

The state of nations is for Hobbes the state of nature. With no common power to hold them in awe, sovereigns are engaged continually in a war of all against all. Not, of course, that there is continual fighting, for 'the nature of war, consisteth not in actual fighting; but in the known disposition thereto, during all the time there is no assurance to the contrary' (*E.W.* iii, p. 113). Hobbes would have approved our phrase *cold war*; it expresses well what he took to be the permanent relationship of nations.

Hobbes recognizes that the state of nations is more tolerable than the condition of individual men in mere nature. Indeed he notes that because sovereigns, by their preparations for war, 'uphold thereby, the industry of their subjects; there does not follow from it, that misery, which accompanies the liberty of particular men' (*E.W.* iii, p. 115).

But more generally the state of nations proves tolerable because it lacks the fundamental equality Hobbes finds in the state of nature—that 'the weakest has strength enough to kill the strongest' (*E.W.* iii, p. 110). From this equality war continually arises. But in the time of Hobbes nations could not kill one another. If we assume that the death of a nation entails not just the death of its particular sovereign, but the destruction of its system of sovereignty, and thus the breakdown of the entire social fabric sustained by that sovereignty, then death has been a relatively rare event in the world of sovereign powers.

The advent of nuclear weapons is, however, bringing the state of nations nearer to the true Hobbesian state of nature. Today the major nuclear powers share the equality of Hobbesian men—they can utterly destroy one another. If nuclear weapons continue to spread, and if nuclear technology continues its present rate of advance, then an ever increasing number of nations will come to share this dreadful equality. And so we may look to Hobbes's account of the natural condition of mankind, with a view to understanding better our own international situation.

The interests and values of nations, like those of Hobbesian men, are essentially subjective and selfish. Governments are pledged to uphold the well-being of their citizens, and can consider the interests of other peoples only as means to their own internal objectives. Foreign aid is good if it increases one's security, gains one access to needed raw materials, or opens up new markets; its effect on those aided is incidental.

From these essentially selfish interests and values, the Hobbesian causes of war can easily be seen to follow. The root cause again is competition—the existence of incompatible national aims and objectives. The conditions for the well-being of one people are, or at least appear, incompatible with the conditions for the well-being of another.

Competition gives rise to diffidence. Fear develops that the potential enemy is an actual enemy, planning to forestall competition by a sudden resort to military conquest. And this fear converts potential enmity into actual enmity; once it takes hold, no real grounds of competition are needed to sustain a cold war. Diffidence is self-perpetuating.

And finally glory—the desire of nations to have their own subjective valuations accepted by others, the desire not to lose face—operates unmistakably on the international scene. From the building of the Berlin Wall to the bombing of Hanoi, its effects rampage around the globe.

Lack of common interests and presence of conflicting individual interests thus give rise to continual cold war—which in the nuclear age entails the constant fear of violent death. Each nation seeks to preserve and strengthen itself, seeks 'power after power', and becomes progressively more impotent. Each new effort we undertake to increase our security merely increases the insecurity of others, and thus leads them to new efforts which reciprocally increase our own insecurity. This is the natural history of an arms race—a history which bids fair to conclude, later if not sooner, in mutual annihilation.

Fear of nuclear death provides the common interest which alone can provide nations with a basis for common action. As long as merely private interests hold sway, as long 'as private appetite is the measure of good, and evil', nations remain in a condition of war; thus all agree 'that peace is good, and therefore also the way, or means of peace' (E.W. iii, p. 146).

But in the condition of nature peace cannot be had merely for the asking. Nations may agree that hostility is bad, but the causes of hostility—competition, diffidence, and glory—are not terminated by this agreement. Nations find themselves wishing to be out of the cold war, but nevertheless in it, and they must adopt a policy which reflects both their aims and their situation.

Hobbes presents the basis of this policy in his conflation of the first laws of nature with the right of nature: '. . . it is a precept, or general rule of reason, *that every man* [nation], *ought to endeavour peace, as far as he* [it] *has hope of obtaining it; and when he* [it] *cannot obtain it, that he* [it] *may seek, and use, all helps, and advantages of war*' (*E.W.* iii, p. 117).

The right of nature provides a short-term policy; the law of nature a long-term policy. The short-term policy, to be used while the cold war continues, is *deterrence*—to use one's power and strength, as best a nation can, to defend itself, to forestall its own destruction by providing itself with the capacity to reply in kind to any attack. The long-term policy is *disarmament*—to give up those rights which are incompatible with a condition of secure peace, and thus to give up the right to arms. Disarmament will lead nations out of the cold war, but it is a policy which must be undertaken mutually, or not at all.

Hobbes would be critical of two points of view frequently advanced in recent years. He would disagree with those who argue that in virtue of the peril of war, we must concentrate all our efforts on peace, refraining from any belligerent or hostile action even in what we might consider our own defence. He would equally disagree with those who argue that in virtue of the peril of communist or capitalist domination, we must concentrate all our efforts on preparing for war, refraining from any peaceful overture or initiative, for fear of appearing weak or unwilling to fight. Hobbes we may suppose, would have approved of the actions of the late President Kennedy, during and after the 1962 Cuban crisis.

Hobbes's opposition to unilateral renunciations of right and power is indeed quite explicit. To paraphrase his commentary on the second law of nature, as it might be applied here: as long as every nation holds the right to maintain arms, and to use them as it likes, so long are all nations in the condition of war. But if other nations will not lay down their right to arms, then there is no reason for some one nation to divest itself of this right, for that would be to expose itself to attack, which no nation is bound to do, rather than to dispose itself to peace. (Cf. *E.W.* iii, p. 118.)

But the second law itself may be paraphrased as a plea for multilateral disarmament: that a nation be willing, when others are so too, as far-forth as it shall think necessary for peace, to lay down its right to maintain arms, and be contented with so much power against other nations, as it would allow other nations against itself. (Cf. Ibid.)

It is undoubtedly difficult, physically and psychologically, to pursue conjointly the policies of deterrence and disarmament. It is hard to seek peace but also to prepare for war. But the situation in which nations find themselves is hard, and Hobbes would not suppose it to be easily altered. The condition of men in nature is precarious and perilous; if they are able to escape from the state of nature, they are fortunate. The same is true of nations.

Nations then must agree to disarm while maintaining their guard against each other. But how is this agreement to be effected? How is the common interest to be made effective against the various particular interests which may lead men to take advantage of, or to ignore, the basis of agreement? Remember that we illustrated the problem of making covenants effective in the state of nature, by the example of a disarmament agreement. Hobbes insists:

> And covenants, without the sword, are but words, and of no strength to secure a man at all. Therefore notwithstanding the laws of nature, . . . if there be no power erected, or not great enough for our security; every man will, and may lawfully rely on his own strength and art, for caution against all other men. (*E.W.* iii, p. 154)

This is as applicable to nations as to individual men. Agreement requires enforcement; enforcement requires power.

Hobbes's sovereign ends the state of nature, and the war of all against all, by erecting a common set of standards as the conditions of peace, and by enforcing them to obtain peace. By analogy, then, we may suppose that Hobbes would consider world government necessary to end the cold war among nations, by establishing and enforcing the conditions of peace. A single will is required to translate the vaguely conceived common interest in peace into determinate and effective actions to secure peace. And this will must have the power to make these actions in the interest of every nation.

Such a world government would not be the limited, democratic meeting of reasonable minds envisaged by such enthusiasts as the World Federalists. The Hobbesian sovereign is and must be an absolute ruler. Now we saw earlier that Hobbes makes it impossible to erect the sovereign, because he cannot provide for the requisite concentration of power. Only if men are more tractable than Hobbes admits, is civil society possible, and if they are more tractable, then absolute sovereignty is unnecessary. Therefore it may be supposed that our attempt to apply Hobbesian standards to the world situation fails. If nations are as intractable as Hobbesian men, there is no hope of world authority; if they are more tractable, then world authority may come about on terms other than those laid down by Hobbes.

This objection is not without point. If nations were in all respects Hobbesian men there would be no way of ending the cold war—except by turning it into a hot war. And if they were quite unlike Hobbesian men, there would be no cold war to end. But it seems to me that, avoiding these two extremes, we must admit that nations are all too similar to Hobbesian men, without supposing their natures to be identical to those of the merely self-maintaining machines described by Hobbes. And this makes Hobbes's arguments relevant, if not literally applicable, to the international situation.

Let us suppose then that nations have sufficient rationality to recognize their present impasse, and to understand the need for international authority to overcome it. Let us suppose them then to be willing—if not *very* willing—to agree to such an authority, and to have sufficient self-control to carry out their agreement.

They will not create an independent and absolute sovereign. Rather, the stronger nations will find it to their advantage to agree to a joint exercise of power—to maintain peace by collective action without sacrificing more of their own superior position than necessary. Thus an 'aristocratic' sovereign will be set up, resembling the Security Council of the present United Nations, but with greatly increased rights and powers, and without the present right of veto by individual members.

Such a world authority represents a possible Hobbesian type of government, although without the unlimited rights and powers Hobbes advocates. For we may expect that the weaker nations will collectively oppose giving this new Security Council greater authority than needed. But in so far as all, or at least most, nations can be brought to recognize the needs for world government, and are prepared to offer it limited voluntary support, it may possess sufficient right and power to overcome the present state of cold war, and provide a secure and peaceful basis for world development.

Thus we may hope to avoid Hobbes's absolute sovereign, if we cannot expect the limited democracy of the World Federalists. No doubt world government will be a danger to the nations over which it rules; no doubt it may become a world tyranny. But as Hobbes says: 'And though of so unlimited a power, men may fancy many evil consequences, yet the consequences of the want of it, which is perpetual war . . ., are much worse. The condition of man in this life shall never be without inconveniences; . . .' (*E.W.* iii, p. 195).

Hobbes states a fundamentally important principle in this argument —a principle which many persons, whether sympathetic to or critical of steps toward national disarmament and international authority, overlook. The principle is this: the alternative to an intolerable situation is never an ideal situation, but rather a barely tolerable situation. We must replace presently intolerable risks by tolerable ones; to object that we shall still be running risks is foolish and irrelevant; to reply that there will be no risk is either naïve or dishonest.

It is neither naïve nor dishonest to hope that in time these risks will lessen. Hobbes's men never acquire any genuine regard for one another; they remain always potential enemies, held in harness by the power of the sovereign. But real men do become sociable and genuinely moral.

The child who has no thought for anything but the satisfaction of his present desire becomes the adult who can display genuine concern for the well-being of others. The first step is no doubt to acquire a prudential regard for others, recognizing the need to consider them in order to be considered in return. This is moral adolescence; from prudential regard the next step is to a genuine moral regard, which is independent of the individual's own interests and well-being. If no one attains full moral adulthood, yet few remain completely in moral childhood or adolescence.

Similarly, we may hope that nations will come to display genuine concern for one another. From their present childish stage of selfish competition they must pass to the adolescent stage of grudging co-operation, held together by the power of world government and the fear of nuclear war. But from this they may proceed to become willing partners in a world order dedicated to the equal well-being of all peoples. If this moral adulthood of mankind, like the moral adulthood of the individual, is doomed to imperfect realization, yet mankind is not thereby relegated permanently to its present childhood.

Hobbes can improve our understanding of international affairs. But here, as elsewhere, he is but a limited guide. His picture of mankind shows an important, but partial truth. It is a truth which we must endeavour to overcome—but we shall not overcome it if we deny it, or if we ignore it.

INDEX